D1175407

THE UFO INVASION

THE UFO INVASION

The Roswell Incident, Alien Abductions, and Government Coverups

Edited by Kendrick Frazier, Barry Karr, and Joe Nickell

59 John Glenn Drive
Amherst, New York 14228-2197

Published 1997 by Prometheus Books

01 00 99 98 97   5 4 3 2

Library of Congress Cataloging-in-Publication Data

The UFO invasion : the Roswell incident, alien abductions, and government coverups / edited by Kendrick Frazier, Barry Karr, and Joe Nickell.
p.   cm.
Includes bibliographical references.
ISBN 1–57392–131–9 (cloth : alk. paper)
1. Human-alien encounters. 2. Unidentified flying objects—Sightings and encounters—New Mexico—Roswell. 3. Unidentified flying objects—Sightings and encounters. 4. Alien abduction. I. Frazier, Kendrick. II. Karr, Barry. III.Nickell, Joe.
BL2050.U36   1997
001.942—dc21   96–53067
CIP

Printed in the United States of America on acid-free paper

CONTENTS

PART THREE: ROSWELL AND THE "ALIEN AUTOPSY"

PART FOUR: OTHER UFO CASES

PART SIX: CROP CIRCLES

PART SEVEN: EXTRATERRESTRIAL INTELLIGENCE

INTRODUCTION
Kendrick Frazier

Our fascination with UFOs goes to the heart of one of the seminal questions of human existence: Are we alone in the universe?

We don't yet know the answer. Until we do, we will continue the quest for some kind of resolution. And until it is resolved, if ever, the search will be conducted in every way humans find possible, from the very scientific and scholarly to the highly speculative and anecdotal, and every permutation in between.

The modern UFO movement began on June 24, 1947. A pilot named Kenneth Arnold flying in a private plane toward Mt. Rainier in Washington State reported a bright flash and then what seemed to him to be nine disk- or saucer-shaped objects. They seemed to be flying in a chainlike formation five miles long and to dip or change direction.

Arnold later that day took his sighting to a local FBI office, but it was closed and so he went to the local newspaper in Pendleton, Oregon, and related his story to a columnist there, saying the objects "flew like a saucer would if skipped over water." The next day the news then went out nationally over the Associated Press: "Nine bright saucer-like objects flying at 'incredible speed' at 10,000 feet altitude were reported here today by Kenneth Arnold, Boise, Idaho, a pilot who said he could not hazard a guess as to what they were."

The story appeared worldwide. Flying saucers were suddenly all the rage. More sightings came in. People were seeing lights in the sky and "flying saucers" almost everywhere. Some thought they were secret U.S. or Soviet aircraft being tested. Others, that they were spaceships from another planet.

It was in this charged atmosphere that a New Mexico rancher named William W. "Mac" Brazel innocently became a player in this drama. On June 14 while riding the range on his ranch thirty miles southeast of Corona, New Mexico, Brazel had found some unusual-looking debris. It was, he said, mostly pieces of paper covered with foil, small sticks, and torn pieces of gray rubber, scattered over an area of about two hundred yards. The largest of the pieces were only about three feet across. It looked kind of like a kite but clearly wasn't a kite. He didn't know what it was. On July 4 he and his family went back to the site and recovered all the pieces they could find and took them back to their ranch house.

Brazel hadn't heard about the "flying disc" reports, but his brother-in-law had and persuaded him to go to the sheriff's office in Roswell, New Mexico, and report his discovery in case there was some relation. Somewhat sheepishly (and to his immediate regret), on July 7 Brazel did so, and the sheriff referred the discovery to intelligence officers at the nearby Roswell Army Air Field who came out to the ranch and got some of the debris.

Exactly how Brazel's modest-seeming debris got transformed momentarily into the remains of a crashed flying saucer we may never exactly know. Stories differ greatly. Fact and myth quickly diverge.

What we do know is that a base public information officer called a local radio station and announced that the air force base had recovered a "flying disc." This became a one-day sensation as a news story ("RAAF Captures Flying Saucer on Ranch in Roswell Area," was the headline on the July 8 *Roswell Daily Record*), until the news was retracted the next day when the Eighth Air Force at Forth Worth, Texas, announced that the debris in fact consisted of parts of angular corner radar reflectors suspended from balloons. The launches were for training and experimental purposes. That explanation was widely reported in newspapers of July 9. Local newspaper reporters and photographers were even brought out to Alamogordo Army Air Base on July 9 and shown foil-covered reflectors being launched from there and other locations ("Fantasy of 'Flying Disc' Is Explained Here" was the front-page headline of the July 10 *Alamogordo News*), and the original story was quickly discounted and forgotten.

Thus, remarkably, within a two-and-a-half-week period in the summer of 1947 three of the major themes that have characterized the UFO movement in the ensuing half century were struck: sightings of fast-moving lights or objects in the sky; reports of recovery of debris of some sort northwest of Roswell; and government involvement and explanations.

Four other themes are of more recent vintage: claims that aliens have made contact with individual humans (contactees), highly popular in the 1970s and 1980s; claims, usually brought out under hypnosis, that certain humans have been abducted by aliens and even taken on board alien spacecraft (abductees), popular in the 1980s and 1990s; claims that the government has recovered aliens from either the Roswell incident or other UFO crashes (something also of fairly recent vintage); and claims that the government has been covering up all these and other matters (a recurring theme since at least the 1960s).

Two more tendencies have become apparent since then. Stories that were once thought dead, perhaps for good reason, can be resurrected years, even decades, later by popular writers and UFO proponents who report them to a new generation of readers. And stories that may once have been considered too outlandish on their face to be taken seriously by mainstream media now are often treated with a newfound respect. This is especially the case with the stories of abductions and recovered aliens.

What is the reality? Is this all a modern myth or is there some semblance of fact here? Are we actually being visited by alien spacecraft even while, paradoxically, first-rate astronomers involved in the scientific search for extraterrestrial intelligence (SETI)—making use of powerful radio telescopes and computers capable of sorting through millions of channels simultaneously—report they have yet to find anything definitive? Do the tireless promoters of UFO sightings, contacts, abductions, crashes, and coverups really know something that the scientific community does not? What does all this mean, anyhow?

This book is intended to help you answer those questions.

In 1976, a new organization of distinguished scientists, scholars, educators, writers, and investigators was created to help the public sort fact from myth and sense from nonsense among the entire range of paranormal, fringe-science, and borderland-science claims and phenomena. Founding Fellows included authors Isaac Asimov and Martin Gardner, astronomer Carl Sagan, psychologist B. F. Skinner, philosopher Paul Kurtz (the founding chairman), and a number of other well-known philosophers, psychologists, and physical scientists. The nonprofit (and very independent) Committee for the Scientific Investigation of Claims of the Paranormal (CSICOP) seeks first—through networks of scientists, scholars, and investigators who carry out the work—to *investigate* and find answers to questions that excite the popular imagination or have significance for scholarly inquiry. It seeks second to *educate* the public about the results of those inquiries. More generally, it encourages an attitude of critical thinking and responsible, tentative skepticism toward all new claims and assertions, no matter the subject. Whatever the topic, it attempts to bring the full arsenal of science and reason to its inquiries. CSICOP has received wide acclaim from the scientific and scholarly communities and considerable attention from the news media for its work. While the organization has not eschewed controversy, it has always attempted to bring clear-headed analysis and fair-minded evaluation to the claims it considers.

CSICOP's bimonthly journal, the *Skeptical Inquirer,* subtitled The Magazine for Science and Reason, has been the main forum for publication of the results of these inquiries. Reports of UFO sightings, crashes, abductions, and coverups have been regular and frequent topics of investigation in the *Skeptical Inquirer.*

The UFO Invasion is a unique compilation of the major investigations and inquiries on these subjects reported in the pages of the *Skeptical Inquirer.* The "UFO invasion" of the title does not necessarily imply that we have in fact been, or are being, invaded by alien spacecraft. But for the past half century we have definitely been invaded by reports, claims, and assertions of UFOs, disseminated widely by the electronic and print media, too often with little critical analysis. What they actually are, what they actually represent, is the subject of this book.

Three multisubject anthologies of *Skeptical Inquirer* articles have been published previously (also by Prometheus Books), but this is the first devoted to a single topic. It is organized into seven sections: The UFO Enigma, The Crash at Roswell, Roswell and the "Alien Autopsy," Other UFO Cases, Alien Abductions, Crop Circles, and Extraterrestrial Intelligence. Within each section the articles are organized chronologically so you can get some sense of how a given issue has developed over time.

To ensure completeness and timeliness, each author was given an opportunity to amend or update his or her original contribution with the latest information and perspective. Much of the information presented here has never before been available within the covers of a single book. Very little of it has been available at all in most bookstores. If what you have been hearing about these subjects comes mostly from the mass media and the literature of the UFO subculture, you may find much of what is reported here surprising. We hope you will find it instructive as well—and entertaining.

Part One
THE UFO ENIGMA

1.
THE AIRSHIP HYSTERIA OF 1896–97

Robert E. Bartholomew

During the "Great Airship Wave" in the United States between November 1896 and May 1897, thousands of Americans claimed to have observed an airship. This vessel was typically described as cigar-shaped, having wings and/or propellers and an attached undercarriage; yet, in terms of historical context, the nineteenth century lacked the technological sophistication to successfully fly heavier-than-air machines (Sanarov 1981, 164; Klass 1976, 302). The Wright Brothers did not fly until 1903, and attempts at earlier heavier-than-air flight were crude and erratic at best. According to British aviation historian Charles Gibbs-Smith (Clark and Coleman 1975, 133):

> Speaking as an aeronautical historian who specializes in the period before 1910, I can say with certainty that the only airborne vehicles, carrying passengers, which could possibly have been seen anywhere in North America . . . were free-flying spherical balloons, and it is highly unlikely for these to be mistaken for anything else. No form of dirigible (i.e., a gasbag propelled by an airscrew) or heavier-than-air flying machine was flying—or indeed could fly—at this time.

SOCIOCULTURAL PERCEPTIONS

During the period of the outbreak, although speculation about the stimulus for the sightings varied from misperceptions of natural or manmade

This article originally appeared in the *Skeptical Inquirer* 14, no. 2 (Winter 1990). Reprinted with permission.

bodies (i.e., heavenly bodies or fire balloons) to hoaxes, hallucinations, and so on, the overwhelming belief existed that an inventor had secretly developed the first practical airship.

In terms of sociopsychological expectations of the era, most Americans possessed at least a general idea of how an airship and its occupants should appear. This conception was shaped by the popular literature of the time, which contained large volumes of stories on the sensational, and thus highly marketable, subject of attempts at early flight.

Aerial flight was very much in the public eye just prior to the wave. In 1895, the Swedish explorer Solomon August Andrée made headlines describing plans for an Arctic balloon trip, which he unsuccessfully attempted in 1896, just two months before the outbreak. Andrée died in a second attempt the following year. On May 6, 1896, Samuel Pierpont Langley, described by Gibbs-Smith as "the first major aeronautical figure in the United States," made headlines after successfully testing in flight his large aeroplane model no. 5. About one month before the outbreak, the *New York Times* (September 28, 1896) carried an article with front-page headlines describing the crash of the experimental airship *Albatross*: Inventor/navigator William Paul narrowly escaped serious injury after his craft "dropped rapidly, beat into a clump of trees, and fell." The article concludes: "The inventor says the experiment was unsuccessful because of the quartering northeast wind, and that but for this he would have made a flight to astonish the world."

Further, intense interest in the invention of mechanical contrivances, especially air machines, developed in the early 1890s and resulted in a major weekly series beginning in 1892 that achieved widespread readership (Clarke 1986, 589).

The sightings occurred in two separate waves: the first from November 17 to mid-December 1896, and the second, January 22 to late May 1897 (Bullard 1982a, 207, 211).

Sensationalistic "yellow journalism" typified the period just prior to and encompassing the sightings as newspapers often reported highly speculative stories (or in some case even made up stories) on a wide range of events. One purpose was to create news on "slow news days" in order to increase circulation (Hiebert, Bohn, and Ungurait 1982). One story in particular generated a tremendous amount of newspaper and magazine coverage speculating about the identity of an apparently fictitious airship inventor said to have been constructing such a craft. Whatever the editors' motivation, on November 1, 1896, the *Detroit Free Press* reported that in the near future a New York inventor would construct and fly an "aerial torpedo boat." Sixteen days later, the *Sacramento* (California) *Bee,* printed a telegram from a New York man claiming he and two friends would board an airship of his invention and fly to California, which he promised to reach within two days. Coincidentally, that night the first sightings in the 1896–97 wave were recorded as hundreds of witnesses in Sacramento reported sighting an airship.

This report, and the ones to follow, seemed to spark a snowball effect. Speculative stories about the possible existence of an airship and inventor(s), in addition to reports of other sightings, appeared in hundreds of newspapers and in nearly every state. Based on a collection by T. E. Bullard (1982b) of more than 1,000 separate airship-related newspaper stories from this period, a conservative estimate of the number of alleged individual sightings would be 100,000, as several sightings were said to have involved participation by entire cities and towns.[2] Bartholomew (1989) has analyzed newspaper accounts of witnesses during the wave who (usually alone, at night in isolated areas), similar to those in modern UFO waves, claimed to have conversed with the pilots. However, unlike modern-day encounters, witnesses described occupants "who appeared to be ordinary American citizens and claimed that their invention was about to revolutionize travel and transportation" (Sachs 1980, 9).

LITERATURE SURVEY

A survey of mass-hysteria literature reveals the importance of three key elements in the composition of any case: ambiguity, anxiety, and a redefinition of the situation from the general to the specific. Hall (1972, 216) summarizes the role of these elements:

> The recipe for this type of hysterical outbreak is a combination of a high level of anxiety or tension with some kind of ambiguous event which is interpreted as posing a serious threat. The ambiguous event is transformed, in beliefs, into an unambiguously threatening event which apparently justifies the diffuse anxiety which was its antecedent.

Hall, a UFO proponent, finds fault with the suggestion that many UFO reports (past or present) are due to hysterical contagion. One of his central arguments is that UFO witnesses often fail to interpret the incidents as serious personal threats. Thus witnesses are frequently excited but not scared during an incident. I will argue that contagion can occur in situations where the actual hysterical belief is nonthreatening. The 1896–97 airship wave is viewed as a case of collective wish-fulfillment as a response to rapid sociotechnological strains and to rumors that someone had invented the world's first practical airship.

GENERALIZED BELIEF

In the years leading up to and immediately prior to the airship sightings, the possibility that someone would soon perfect the first practical heavier-than-air flying machine was the subject of widespread speculation in science-fiction stories. This was given special emphasis as the twentieth century approached. In the 1890s, Americans were obsessed with science and inventions. According to Clarke (1986, 589):

> The Frank Reade Library [was] . . . designed to meet the insatiable demand for tales of mechanical novelty by concentrating on a nonstop run of invention stories. The series opened on 24 September 1892 and continued for 191 issues. It was the first serial publication of any size ever to be devoted exclusively to science fiction stories; and every issue throbbed with the dynamism of coming things—robots, submarines, flying machines . . . and the rest of the imaginative bric-a-brac of an age that was in love with the great wonders of science.

Bullard (1982a, 203) also notes that from about 1880 through the early twentieth century widespread publicity in books and magazines helped to mold a common belief that a heavier-than-air vessel would be perfected imminently:

> Magazines devoted to science and engineering vied with Jules Verne's *Robur the Conqueror* and other fictional publications to describe the flier which would soon succeed, and this literature fed the public a steady diet of aeronautical speculation and news to prime people for the day when the riddle of aerial navigation finally would receive a solution.

Further fueling this generalized belief were the growing number of failed aerial trials making news. Although all were unsuccessful in perfecting a practical airship, during "the late 1890s numerous inventors in the United States obtained patents for planned airships" (Brookesmith 1984, 107; Jacobs 1976, 27).

AMBIGUITY

The boom in airship patents during the latter 1890s coincided with the airship wave. (For actual reproductions of some of the original patents, see Lore and Deneault 1968, 16–17, 38–39). Intense competition to be the first to patent such a machine resulted in a shroud of secrecy, as many inventors often withheld vital data on their patents and experimental craft.

As noted in Brookesmith (1984, 107), the air of mystery surrounding the state of aerial development only fostered public belief that a practical airship had been developed.

This view is supported by historian David M. Jacobs (1976, 27–28):

> In the late 1890s many people in the United States obtained patents for proposed airships. Most people believed someone would soon invent a flying machine, and many wanted to capitalize on the fame and fortune that would certainly come to the first person to launch an American into the skies. As soon as someone had a glimmer of an airship design, he immediately applied for a patent. These would-be inventors constantly worried over possible theft or plagiarism . . . [and] most people kept their patents secret. Given this atmosphere and the numerous European and American experiments with flight, it is not surprising that secret inventor stories so captured the public imagination and seemed such a logical explanation for the airship mystery.

Environmental factors further contributed to ambiguity during the episode. As there were a minimum of several thousand sightings, a specific breakdown of each case is unfeasible. However, Bullard (1984, personal communication), commenting on the approximately 1,000 newspaper stories detailing sightings that he had collected during the wave, noted that approximately 80 to 90 percent of the cases were reported to have occurred at night. Other researchers have noted the overwhelming tendency of the airships to appear at night (Berliner 1978, 2; Sanarov 1981, 166). Also, the wave occurred primarily during the winter months and abruptly ended in early spring, coinciding with a reduction in hours of sunlight.

Further inducing ambiguity were the mysteries associated with the airship. Who actually was the inventor? How had he accomplished this great feat? Who helped him, if anyone? Where was his secret hideout? Where would he test his machine next?

ANXIETY AND INTENSE EXCITEMENT

The wave occurred during a period of rapid technological change and amid intense public interest in airship development. As detailed earlier, a widespread belief circulated in the United States just prior to the outbreak that someone had invented the world's first practical airship. A major role in spreading this belief was played by period newspapers, characterized by sensationalism and intense speculation on issues of the day. Newspaper publisher William Randolph Hearst noted this in an editorial attacking such press coverage:

> "Fake journalism" has a good deal to answer for, but we do not recall a more discernible exploit in that line than the persistent attempt to make the public believe that the air in this vicinity is populated with airships. It has been manifest for weeks that the whole airship story is pure myth. (Klass 1976, 314, citing *San Francisco Examiner,* December 5, 1896)

Bullard (1982a, 224) and Klass (1976, 314–15) also concur with the belief that newspapers exerted considerable influence in perpetuating and maintaining the outbreak.

A. M. Herring, writing in the *Scientific American* of June 26, 1897, noted the intense experimentation and the widespread publicity of the belief that a practical airship existed in the late 1890s, but "especially" in the period of time coinciding with the airship outbreak:

> This line of experiment has resulted in such great progress in the last few years (and especially so in the last six months) that attainment of long, free flight for man, which not long ago seemed an invention for the far distant future, is a thing now near, if not quite at hand. (403)

Neeley (1979, 68) attributes the episode to social stress fostered, in part, by rapid technological changes. Neeley surveyed 223 Illinois newspapers during the outbreak. He clearly applies his Illinois findings to the larger pattern of reports across the United States:

> Let us first consider the people of 1897. They lived in very interesting and stress-filled times. They were amazed at the technological achievements of the time. The telephone was merely fourteen years old, electricity had just been made available for practical uses, x-rays had been discovered merely two years earlier. The horseless carriage was just around the corner as was flight. They had just dealt with a bad winter and spring had brought forth one of the greatest floods to hit the Midwest. It was raining constantly and only snow broke the monotony. A clear sky was a rarity. Affairs had just returned to normal following the Civil War and there were accounts of wars in Greece and Cuba. . . . Jules Verne was writing stories of . . . an electric airship. Suddenly the skies clear and in the northwest a bright light was seen. The cry "Airship!" went up and a crowd gathered to watch. Soon a cloud obscured it and the airship had "left." Or a bright light was seen in the southeast and the witnesses "followed" its path behind a cloud until a bright light was seen in the northwest. Surely they had seen the airship cross the sky.

REDEFINITION OF THE SITUATION

The airship wave occurred in two separate phases: the first primarily between November 17 and mid-December 1896, and the second between January 22 and late May 1897. The separate waves closely paralleled newspaper accounts of where the airship would appear. For instance, the overwhelming majority of sightings in the 1896 wave took place in California, and all of the sightings occurred within the general Pacific Coast region (Bullard 1982b). From a definitional view it's interesting that the popular belief prior to and during the November-December 1896 wave held that an inventor would fly an airship to California and then slowly progress back across the country, ending in New York. The popular newspaper accounts circulating during the second wave (although there were a variety of stories) centered around an inventor partaking in a transcontinental airship flight. One story told how the inventor would fly his airship across the country to Washington, D.C., where he would take out a patent. Another speculated that the United States government was secretly testing an airship by flying it across the country. Coincidentally, the second wave began in the western United States and worked its way eastward in an erratic but systematic pattern, so that the 1897 wave closed abruptly in early May with sightings on the coastal northeast:

> Suddenly the climax. The conclusion to the extraordinary transcontinental voyage was reached. On April 30, 1897, the great airship was seen over Yonkers, New York . . . at 3 A.M. . . . toward the sea.
>
> . . . Curiously, when the 1896–97 complex stopped, for all practical purposes it stopped cold. Various sightings continued to be recorded through the years, but this particular phenomenon reached a dead end at the shores of the Atlantic. . . . Virtually no new sightings emerged from the areas over which it had soared. It was all over. (Flammonde 1977, 115–17)

During both waves, the cultural expectation of the time appears to have been shaped and defined by newspaper accounts and subsequently fulfilled by the pattern of reports. It appeared that the collective consciousness, as reflected and defined in newspaper stories, created a consensual belief that the airship had completed its transcontinental flight. This would explain not only the general west-to-east pattern across the country but also the abrupt end to the wave.

A survey of the more than 1,000 original airship reports from United States newspapers collected by Bullard (1982b) shows that most sightings of unidentified aerial objects between November 17, 1896, and late

May 1897 closely paralleled popular literature accounts of early heavier-than-air travel attempts. An examination of Bullard's data shows that whenever specific descriptions of airships were given, beyond the interpretation of ambiguous nocturnal aerial lights, eyewitness accounts vacillated between two types of craft. One was a large oblong or egg-shaped main structure having wings similar to those of a bird. These wings were frequently reported to be "flopping" in a birdlike manner. The second craft type also consisted of a large central portion, but sported propellers or fanlike wheels. Both types of craft were said to possess powerful searchlights and some type of motor propulsion system, and often had a carriage suspended under the main structure. The drawing in figure 1 is of an airship reported by hundreds of persons on November 23, 1896, over the city of San Francisco. The description conforms to cultural expectations of how an American citizen of 1896 would project such a craft to appear. None of the vessels were described in terms of more contemporary disc or saucer shapes. Other sightings during the wave resembled a common type of UFO description. (See figure 2.)

These descriptions closely mimic early heavier-than-air flight attempts. For instance, the first known manned, powered flight was Heneri Giffard's steam airship (figure 3). The large cigar-shaped top portion, with a smaller basket underneath, featured a structural design commonly reported 45 years later during the 1896–97 U.S. airship wave.

Figure 4 shows a model of the first airship to complete a circular flight. On August 9, 1884, the *La France* flew nearly five miles at an

Figure 1. From the *Skeptical Inquirer.* An artist's rendition of the airship reported seen by hundreds of people over San Francisco on November 23, 1896. (Source: *San Francisco Call,* November 23, 1896, p. 1.)

average speed of thirteen miles per hour. A very similar type of airship was reported on April 10, 1897, over the city of Chicago. Grabbing his son's box camera, Walter McCann claimed to have taken two photographs. An etching of the best photo, appearing on the front page of the *Chicago Tribune* of April 12, is depicted in figure 5. The picture was taken as the craft allegedly sailed over a suburb at approximately 6 A.M. The pictures were taken during the height of a monthlong airship wave in Illinois, with thousands of reported sightings.

Figure 2. From the *Skeptical Inquirer.* Airship sighted over Oakland, California, between November 17 and 19, 1896. (Source: *San Francisco Call,* November 19, 1896, p. 1.)

CONCLUSION

In the presence of the widespread airship rumors holding that such an invention was on the verge of perfection, the ambiguity of the nighttime sky, and the intense emotions held by many Americans that such a dramatic achievement was at hand—and the fanning of these emotions by speculative and often fabricated newspaper stories—people attempted to relieve their emotionally aroused states by looking to the skies for proof or disconfirmation of the airship-invention stories. They expected to see airships and saw them. Whereas contemporary people collectively perceive "flying saucers" from outer space, citizens in 1896–97 were predisposed by popular literature of the era to see airships. Research on autokinetic movement appears applicable, as it concerns problem-solving dynamics (Turner and Kirlian 1972, 35). Interpretation of ambiguous stimuli within a group setting will result in members' developing an increased need to define the situation, depending less on their own judgment for reality validation and more on the judgment of others (reality testing).

> When the stimulus situation lacks objective structure, the effect of the other's judgement is . . . pronounced. . . . In one . . . study of social factors in perception utilizing the autokinetic phenomenon, an individual

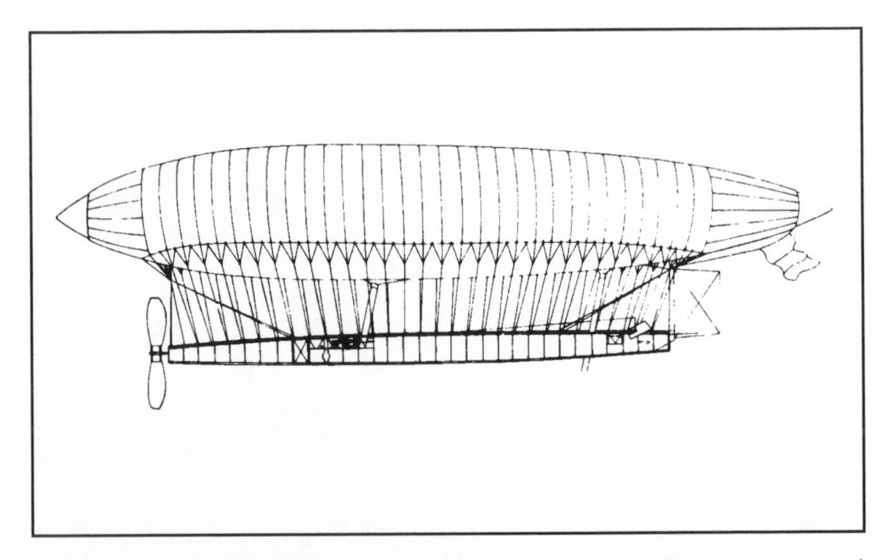

Figure 3. From the *Skeptical Inquirer.* Heneri Giffard's 1852 steam-powered airship. (Source: B. Collier, *The Airship: A History,* Hart-Davis, MacGibbon, London, 1974, p. 29.)

Figure 4. From the *Skeptical Inquirer.* The *La France* circa 1884. (Source: C. H. Gibbs-Smith, *Flight Throughout the Ages,* Thomas Y. Crowell, New York, 1974, p. 76.)

Figure 5. From the *Skeptical Inquirer.* Walter McCann's alleged photo of an airship over Chicago. (Source: *Chicago Tribune,* April 12, 1897. p. 1).

> judged distances of apparent movement first alone and then with two or three other subjects. This unstructured situation arouses considerable uncertainty. Even though they were not told to agree and were cautioned against being influenced, the individuals in togetherness situations shifted their judgement toward a common standard or norm of judgement. . . . The influence of various individuals differed, and the emerging common norm for judgement was in various instances above or below the average of individual judgements in the initial session alone. (Sherif and Harvey 1952, 302)

Research on the "autokinetic effect" is of more specific interest, as it has shown that individual judgments tend to agree in a group setting while observing the common stimulus of a pinpoint of light within a dark environment. This effect is well known in social psychology and was first demonstrated by Sherif (1936). Individuals in situations lacking in stable perceptual anchorages begin to feel a sense of uneasiness, with anxiety generated as the person experiences a heightened need to visually define or make sense of the light. In group settings, individuals will attempt to reduce the anxieties created by an uncertain situation. Beeson (1979, 180) outlines this process:

> A viewer in a completely dark room seeing one pinpoint of light experiences a visual stimulus without its normal attendant visual context. Up, down, back, forward, far and near, exist in relation to other stimuli and when this frame of reference is missing, the light is free to roam in one's perceptual field. It is for this reason that considerable random motion will be experienced by anyone viewing the light.

Within highly ambiguous situations, such as the people scanning the nighttime skies for an imaginary airship, "inference can perform the function of perception by filling in missing information in instances where perception is either inefficient or inadequate" (Massad, Hubbard, and Newtson 1979). Accordingly, individuals with an airship "mind-set" perceived airships. Today, with the existence of a collective belief in extraterrestrials traversing the skies, usually at night, flying saucers are seen. Allan Hendry, former editor of the *International UFO Reporter,* a scientifically orientated UFO publication, provided a good example of this process. He noted in 1978 that a large number of advertising planes had been initially mistaken for UFOs and were described as having been distinctly disc- or saucer-shaped:

> In the three hundred calls that . . . [our organization] has dealt with that were based on confirmed ad planes at night, 90 percent of the witnesses described not what was perceptually available, but rather that they could see a disc-shaped form rotating with "fixed" lights; many of these people imagine that they see a dome on top and, when pressed, will swear that they can make out the outline with confidence.

Overall, the sightings appear to have functioned as a reassuring symbol during a period of great uncertainty with rapid technological changes at the end of the twentieth century. People had great affection for these technological marvels that were changing social patterns that had existed for thousands of years but were simultaneously concerned with the potential destructive power these machines could hold over their lives.

The airship wave functioned to show man's dominance over the untamed and previously sacred skies, leaving them with the comforting belief that a positive element was in control. In the words of Clark and Coleman (1975, 163):

> Most of them [Americans] saw the craft as a sort of final triumph of technology, and something about which they must surely have entertained ambivalent feelings. All the talk about bombs and aerial machine guns, pointing toward a time when there would be no safety anywhere, must have been disconcerting in the extreme. Moreover, now the heavens had been violated; men had tainted even the domain of angels.

It is important to note that, although social strains generated by rapid technological advancement were especially acute during this period, Americans sighting these phantom craft clearly did not fear them. Airships were seen as a positive influence in reaction to the negative strains brought about by rapid technological advancements in a variety of fields. Hence the redefinition of the ambiguous, mundane, predominately nocturnal aerial stimuli (i.e., stars, planets) functioned to create a reassuring presence.

NOTES

1. I am indebted to T. E. Bullard, folklorist, Indiana University, Bloomington, Indiana, for providing access to original airship data.

2. Any such specific estimate is hazardous. However, this figure seems reasonably accurate as a conservative estimate of the minimum number of participants, based on Bullard's data.

REFERENCES

Bartholomew, R. 1989. *UFOlore: A Social Psychological Study of a Modern Myth in the Making.* Stone Mountain, Ga.: Arcturus.

Beeson, R. 1979. "The Improbable Primate and the Modern Myth." In *The Scientist Looks at Sasquatch II,* edited by G. Krantz and R. Sprague, 166–95. Moscow, Idaho: University Press of Idaho.

Berliner, D. 1978. "The Nineteenth-Century Airship Mystery." *International Fortean Organization Journal,* no. 29 (May–June): 2–6.

Brookesmith, L. 1984. *The Age of the UFO.* London: Orbis.

Bullard, T. E. 1982a. *Mysteries in the Eye of the Beholder: UFOs and Their Correlates as a Folkloric Theme Past and Present.* Doctoral dissertation. Indiana University Folklore Department.

———. 1982b. *The Airship File: A Collection of Texts Concerning Phantom Airships and Other UFOs Gathered from Newspapers and Periodicals Mostly During the Hundred Years Prior to Kenneth Arnold's Sighting.* Unpublished manuscript.

Clark, J., and L. Coleman. 1975. *The Unidentified: Notes Toward Solving the UFO Mystery.* New York: Warner.

Clarke, J. F. 1986. "American Anticipations: The First of the Futurists." *Futures* (August): 584–92.

Flammonde, P. 1977. *UFOs Exist!* New York: Ballantine.

Hall, R. 1972. "Sociological Perspectives on UFO Reports." In *UFOs: A Scientific Debate,* edited by C. Sagan and T. Page. Ithaca, N.Y.: Cornell University Press.

Hiebert, R., T. Bohn, and D. Ungurait. 1982. *Mass Media III.* New York: Longman.

Jacobs, D. 1976. *The UFO Controversy in America.* New York: Signet.

Klass, P. 1976. *UFOs—Explained.* New York: Random House.

Lore, G., and H. Deneault. 1968. *Mysteries of the Skies: UFOs in Perspective.* Englewood Cliffs, N.J.: Prentice-Hall.

Massad, C. M., M. Hubbard, and D. Newtson. 1979. "Selective Perception of Events." *Journal of Experimental Social Psychology* 15: 513–32.

Neeley, R. 1979. "The Airship in Illinois." *Journal of UFO Studies,* o.s., 1, no. 1: 49–69.

Sachs, M. 1980. *The UFO Encyclopedia.* New York: Perigee.

Sanarov, V. 1981. "On the Nature and Origin of Flying Saucers and Little Green Men." *Current Anthropology* (April): 163–67.

Sherif, M. 1936. *The Psychology of Social Norms.* New York: Harper and Row.

Sherif, M., and O. J. Harvey. 1952. "A Study in Ego Functioning: Elimination of Stable Anchorages in Individual and Group Situations." *Sociometry* 15: 302.

Turner, R., and L. Kirlian. 1972. *Collective Behavior.* Englewood Cliffs, N.J.: Prentice-Hall.

2. THE CONDON UFO STUDY

Philip J. Klass

Just twenty years ago, the University of Colorado undertook a very controversial project that was unique in the annals of university research. At the request of the U.S. government, the university agreed to perform a two-year scientific investigation into unidentified flying objects—UFOs. That effort probably brought more fame to the University of Colorado—and certainly more criticism—than any other activity ever to take place on its campus.

The late Edward U. Condon, a world-famous physicist who once had directed the National Bureau of Standards with distinction and who was then a member of the Colorado faculty, reluctantly agreed to head the UFO project. He became a favorite whipping boy of the UFO cultists. Robert Low, another faculty member, who served as the project coordinator and later died in an aircraft accident, was defamed in the national news media on the grounds that he and Condon had plotted to "trick" the American public.

There was indeed a plot to mislead the public. *But Condon and Low were its victims, not its architects.* A small group of "UFO-believers," which included a U.S. congressman, secretly plotted to discredit the Colorado effort. Today we can piece together their covert actions because of fortuitous access to the files of the chief architect of the plot, who later committed suicide.

Let me emphasize that I cannot endorse the Colorado investigation as having been well managed. In my opinion, Condon himself did not play

---

This article originally appeared in the *Skeptical Inquirer* 10, no. 4 (Summer 1986). Reprinted with permission.

a sufficiently active role, and Low had no prior experience in coordinating so complex an investigation. But under the circumstances, I doubt that anyone could have done much better. And those who later plotted to discredit the Colorado effort would surely have done far worse had they directed the program.

But first let us turn back the clock to the spring of 1966 for the benefit of those who are too young to recall the circumstances and for those whose memories have dimmed because of the passage of time.

By the spring of 1966, the U.S. Air Force had been investigating UFO reports for nearly 20 years—more than 10,000 of them. At first there were some within the USAF who suspected that some UFO reports might involve extraterrestrial spacecraft, or possibly Soviet reconnaissance vehicles, perhaps built with advanced technology obtained from German scientists captured at the end of World War II.

But by the early 1950s, having learned that many seemingly mysterious UFO reports had been generated by prosaic trigger-mechanisms such as bright celestial bodies, weather balloons, and meteor-fireballs, the USAF concluded that all UFO reports were explainable in prosaic terms. And so the USAF assigned only low-level personnel—many of them without appropriate training—to investigate UFO reports. Thus it is not surprising that in some instances these relatively unskilled investigators had trouble finding rational explanations and that there remained a small percentage of unexplained cases. For those eager to believe in alien spaceships, these unexplained cases seemed to be proof of their fondest hopes.

Sometimes USAF investigators offered explanations that evoked sharp criticism, and with good reason. For example, in March 1966, UFO reports from university students and others in southern Michigan attracted national attention after the USAF's investigator—astronomer J. Allen Hynek—suggested that some of the reports might be due to swamp-gas. This explanation aroused much ridicule and criticism, even from some members of Congress, including a then obscure Michigan congressman named Gerald Ford.

This criticism of the USAF's handling of the UFO issue and charges that it was guilty of a coverup prompted the USAF's scientific advisory board to recommend that an independent UFO study be conducted by one or more universities. But when well-known scientific institutions like MIT were sounded out, none of them was willing to undertake such an effort.

During the late spring and early summer of 1966, as the USAF struggled to find a prestigious scientist and institution willing to tackle the UFO study, a member of the faculty of the University of Arizona was lobbying to obtain the contract. His name was James F. McDonald, an atmospheric physicist who was respected by his peers. There was, however, a basic obstacle for McDonald. *He already was convinced that some UFOs were extraterrestrial craft.* This clearly made it impossible for him to direct or conduct an independent, unbiased investigation.

Condon, who had earlier served as president of the American Physical Society and of the American Association for the Advancement of Science and was a member of the respected National Academy of Sciences, had recently joined the Colorado faculty to devote his later years to research in atomic physics.

Robert Low, then assistant dean of the Colorado graduate school, was considerably younger than Condon. He recognized the potential pitfalls of the UFO investigation, but he believed that the project offered the university the opportunity to achieve wider recognition as a center of scientific excellence and that it could be a useful stepping stone in his own academic career. Events would prove Low to be wrong on both counts.

After weeks of internal debate over whether to take on the program, the university submitted its formal proposal in early October 1966, and on October 7 the USAF announced that Colorado had been selected. A key proviso in the Colorado proposal, which was included in its contract, was the following: "The work will be conducted under conditions of strictest objectivity *by investigators who, as carefully as can be determined, have no predilections or preconceived positions on the UFO question*" (emphasis added).

This was a vitally important condition, because the Colorado group would sit in scientific judgment, much as a jury in a court of law, whose members must swear that they have no prior view on the issues under trial. I suspect, but have no proof, that this proviso was intended to foreclose any direct participation by McDonald. The University of Colorado subsequently awarded subcontracts to several outside institutions and specialists, but none ever went to McDonald.

However, this key provision was violated immediately when David R. Saunders joined the project as one of its three top scientists—referred to as "principal investigators." Saunders was a member of the university's psychology faculty, but it soon became quite evident that Saunders was at least a "quasi-believer," as one of the younger scientists in the project soon characterized him to his face.

Saunders subsequently acknowledged publicly that he had become interested in UFOs at least several months before the USAF contract was awarded. After reading a best-selling "gee-whiz" type book titled *UFOs—Serious Business,* written by a strongly pro-UFO author, Saunders had visited the headquarters of the National Investigations Committee of Aerial Phenomena (NICAP)—then the nation's largest group of UFO believers—to meet with officials and to join the organization.

It was Saunders who played the key role in getting Condon and Low to agree—using Saunders's own words that "we ought to concentrate on trying to identify and develop cases that might support the extraterrestrial intelligence hypothesis."[1] Saunders also embraced the cornerstone of the UFO-believers' faith that the U.S. government was involved in a UFO coverup. As Saunders later wrote: "Almost from the first day of the Project, I had maintained that a 'government conspiracy' to conceal the 'truth about UFOs' from the public was an even more likely hypothesis than ETI."[2]

Saunders was well aware of the provision against hiring investigators who had any "predilections or preconceived positions on the UFO question" contained in the USAF contract and the university proposal, because Saunders himself had participated in writing that proposal. Yet he did not disqualify himself from participation on those grounds.

Condon and Low tried to establish a good working relationship with NICAP officials from the beginning. NICAP director Donald Keyhoe and assistant director Richard Hall accepted an invitation in the fall of 1966 to visit Boulder to brief project scientists. Following this meeting, NICAP officials offered a curious endorsement of Condon and his associates in the January–February 1967 issue of the NICAP publication, *The UFO Investigator*. The article said: "It probably is fair to say that the scientists on the project range from open-minded skeptics *to moderately convinced 'believers'*—which is as it should be" (emphasis added).

While NICAP officials clearly were pleased to find "moderately convinced believers" on the Colorado project shortly after the investigation began, if their assessment was correct, then the terms of the USAF contract had been violated.

McDonald also was invited to Boulder in the fall of 1966 to brief the Colorado team. He returned again in the summer of 1967 to brief the project scientists on the results of his own investigations into UFO reports from Australia and New Guinea. But, beyond this, Condon and Low did not avail themselves of McDonald's frequent offers to become more directly involved in the investigation.

J. Allen Hynek also was invited to come to Boulder to brief its team on his own views about UFOS. But the then leading experienced UFO-skeptic, the late astronomer Donald Menzel, was never invited to visit Boulder. Nor was I, although I had offered my services as a consultant early in the program.

In June 1967, about halfway through the investigation, the Colorado team was joined by Norman E. Levine, who had just received his Ph.D. in electrical engineering from the University of Arizona. Levine publicly acknowledged that his own interest in UFOs had been sparked by McDonald. Shortly after Levine joined the staff, I happened to talk with him on the telephone when I placed a call to Low, who was out of town. During the discussion, I was shocked to hear Levine express views on UFOs that were remarkably similar to those voiced by McDonald, especially since Levine had just joined the project.

The Colorado game plan was to dispatch a small team of scientists to investigate important UFO-sighting reports. To ensure that these were promptly reported to the project, an "early warning network" of several dozen persons around the nation was created, consisting principally of NICAP's own field investigators. Shortly after Levine joined the project, he was named "secretary" of the field investigators to coordinate their efforts. As a result, Levine and Saunders played key roles in determining which UFO reports deserved field investigation.

On September 4, 1967, nearly a year after the investigation began, Saunders spoke before the American Psychological Association at its annual meeting in Washington, as a member of a panel during a session on UFOS. During the question-and-answer period, Saunders was asked: "What is your opinion of the scientific integrity of the Condon Committee?" Saunders replied: ". . . I would not wish to remain associated with anything but an open and impartial investigation." Considering Saunders's own pro-UFO views and the key role that he and Levine were playing in the project, his answer is hardly surprising. He remained with the project until the following February 7, when he was fired, along with Levine.

Saunders chose not to reveal to the APA meeting that he was somewhat unhappy with Condon and Low because a few weeks earlier they had rejected two of his suggestions. One of these was that the project should not wait to make its final report before "going public" with some of its interim case-investigation results. Saunders wanted to release those that he believed supported the extraterrestrial hypothesis as well as those that did not.

Saunders's second suggestion was that the Colorado project should encourage public discussion "of the social problems that the world would have to face if either our study or some future study were to generate conclusive evidence of extraterrestrial visitations."[4]

The decision by Condon and Low to reject these two suggestions was a very wise one at the time, and certainly in retrospect. Condon, and the project, had been criticized earlier—with good reason—because of several speeches Condon had given early in the effort in which he revealed his strong skepticism about UFOs being extraterrestrial visitors. As a result, the decision had been made to hold back further public discussion until the final report was issued. Saunders himself had violated that policy by agreeing to speak to the APA without obtaining advanced approval to do so.

Were it not for McDonald's tragic suicide in June 1971 and the fact that a young graduate student named Paul McCarthy, at the University of Hawaii, decided to write his Ph.D. thesis on the UFO controversy, the world would never have known of the covert effort spearheaded by McDonald to torpedo the credibility of the Colorado investigation.

Young McCarthy, who honestly characterized himself as being a "UFO-believer," obtained permission from McDonald's widow to gain access to all of his personal papers. Because of McCarthy's strong pro-UFO views and his understandable sympathies for McDonald in the wake of his tragic death, it seems likely that McCarthy left out at least some "incriminating evidence" from his thesis, completed in late 1975. But there is enough to piece together the curious puzzle.

McDonald had himself become such a fanatical believer in UFOs being alien craft and the greatest scientific mystery of this century that at first he was certain the Colorado group would reach the same conclusion. If it did, then almost certainly the government would then authorize a massive followup UFO investigation, which McDonald would be the logical person to direct because of his expertise. Condon and Low indicated no interest in conducting followup studies. But if Condon and Low remained unconvinced and the two project leaders reached a "negative" conclusion, then McDonald's ambitions would be dashed.

From McCarthy's thesis we learn that when McDonald returned from Boulder in early August 1967 after briefing the Colorado team on the results of his own UFO-case investigations in Australia, he was discouraged. McDonald so indicated in his letter of August 11, 1967, to Mary Lou Armstrong, administrative assistant to the project. In McCarthy's thesis he notes: "McDonald could talk and write openly to Armstrong and a few others, although Condon and Low, the leaders, remained at a distance."[5]

It was at about the same time—August 1967—that a member of the project, believed to have been Mrs. Armstrong, discovered a memo that had been written a year earlier by Robert Low. In this memo, dated August 9, 1966, Low expressed his *personal ideas* on whether the university should take on the UFO study. By late April 1968, the contents of this memo would be the cornerstone of a feature article in *Look* magazine charging that the Colorado investigation was "a half-million dollar 'trick' to make Americans believe the Condon committee was conducting an objective investigation."

The *Look* article was written by John G. Fuller, who earlier had authored two pro-UFO books. Later, Fuller wrote a book endorsing a famous South American "psychic surgeon," and still later a book claiming that some Eastern Air Lines jetliners were haunted by the ghosts of members of a flight crew killed in a tragic accident.

As we consider the contents of Low's memo, we should remember that it was written at a time when there was strong disagreement among faculty members over whether the university should undertake the effort. Low's memo, entitled "Some Thoughts on the UFO Project," was addressed to E. James Archer, dean of the graduate school, and to Thurston E. Manning, vice-president for academic affairs.

His memo noted that those who opposed university involvement felt that "to undertake such a project one has to approach it objectively. That is, one has to admit the possibility that such things as UFOs exist. It is not respectable to give serious consideration to such a possibility. Believers, in other words, remain outcasts." But Low quoted another scientist at a separate research center in Boulder, who favored taking on the project, saying: "We must do it right—objectively and critically . . . having the project here would not put us in the category of scientific kooks."

Low's memo drew an analogy with the ESP experiments that had been conducted in the 1930s by J. B. Rhine at Duke University. Low noted that "the Duke study was done by believers who, after they finished, convinced no one. Our study would be conducted almost exclusively by non-believers who, although they couldn't possibly prove a negative result, could and probably would add an impressive body of evidence that there is no reality to the observations."

Then followed the sentence that later would prove so embarrassing. Low wrote: "The *trick* would be, I think, to describe the project so that, to the public, it would appear a totally objective study, but to the scientific community would present the image of a group of nonbelievers trying their best to be objective but having an almost zero expectation of finding a saucer" (emphasis added).

For most Americans, the word trick has a devious meaning, but according to the *Random House Dictionary of the English Language* "trick" also means "the art or knack of doing something skillfully." That is a meaning more often used by the British and members of the British Commonwealth.

For example, in early 1978, Canadian and U.S. scientists were trying to locate radioactive debris from a Soviet satellite that had reentered prematurely over Canada's Northwest Territories. A United Press International article on the incident quoted a Canadian scientist, named Jack Doyle, as saying: "The *trick* at the moment is to convert blips on our [instrument] tapes to something we can see on the ground" (emphasis added). Robert Low had studied at Oxford University and, during our brief meeting in Washington, I noticed that he had acquired some British speech affectations and word usages. But I acknowledge that it is not possible today to know for sure what meaning he intended in his now infamous memo.

"CONDON, THE TRUE SCIENTIST, UNDERSTOOD SELF-DECEPTION"

Useful insights into the controversy that swirled around the University of Colorado UFO study are offered by Lewis M. Branscomb, chief scientist and vice president of IBM, who in the mid-1960s was chairman of the Joint Institute for Laboratory Astrophysics (JILA) on the university's Boulder campus. Branscomb knew well both Edward U. Condon, who directed the UFO study, and Robert Low, the project administrator. A copy of the accompanying article was submitted to Dr. Branscomb for comments and his response included the following:

"I know first hand that Bob Low did indeed use the word 'trick' in the sense you defined. . . . Scientists, American as well as British, frequently use it to mean a clever or ingenious solution to a problem. The word 'scheme' has a similar ambiguity. (That episode taught me a lesson, and I have avoided using 'trick' in that sense.) . . .

"Condon originally requested that the UFO project be undertaken within JILA, and the proposition was debated by the Fellows. We declined, in part because . . . Condon declined to put in place a set of committed procedures and safeguards that we felt would be necessary to preserve the integrity of so controversial a project. . . . Some of us were concerned that he trusted too many people and might well be victimized, as indeed happened. . . .

"I remember, vividly, a long discussion with Ed Condon in his office . . . when he was considering taking the [UFO] project on. He told me he thought the chance that he could find evidence for a UFO of extraterrestrial origin was infinitesimal, a million-to-one shot. 'But,' he said with the gleam in the eye that betrays a true scientist on the track of remarkable discovery, 'if there is a chance, even the most remote chance that there is something there, I want to be the one to discover it.'

"In that sense, he and McDonald shared the same motivation, coming from opposite intellectual traditions. Each of them wanted a shot at the immortality that would come from the most astonishing discovery in human history. Unlike McDonald, however, Condon, the true scientist, understood self-deception. The stronger the incentive to discover, the greater the temptation to let down one's guard," Dr. Branscomb noted.

Philip J. Klass

The important thing, however, is whether Saunders and McDonald were surprised and shocked when they first read the memo and concluded that Low and Condon had resorted to devious skullduggery. According to the book that Saunders later wrote, he was neither shocked nor surprised, because the memo "did *not* say anything new—it merely expressed concisely what we knew anyway on the basis of Low's day-to-day behavior." If Saunders believed at the time that Low's memo offered evidence of skullduggery, he could, and should, have mentioned it several weeks later when he met with vice president Manning to complain because Condon and Low had rejected his two suggestions. *Yet Saunders did not even mention the Low memo to Manning.*

It is not known with certainty when McDonald first learned of the Low memo. Based on his very close relationship with Levine and Mrs. Armstrong, both of whom saw the memo in August 1967, it would be logical to expect that they had promptly informed McDonald.

Around mid-September 1967, Condon spoke on UFOs at the Atomic Spectroscopy Symposium at the National Bureau of Standards. Condon focused on humorous cases, such as those involving persons he had interviewed who claimed to have made contact with UFOnauts, as well as some incidents that had been exposed as hoaxes. McDonald was especially disturbed by Condon's remarks and wrote to Low to complain, but received no reply.

According to McCarthy's thesis, "Condon's National Bureau of Standards talk was apparently a turning point, for after that McDonald met with Saunders and Levine in early November and talked of engineering a confrontation."[7]

On December 12, 1967, Saunders, Levine, and Armstrong met secretly in Denver with McDonald and Hynek. According to Saunders's book, "McDonald's chief interest was the publication of a newsletter which might later grow into a scientific journal. Hynek wanted to create a visible group of qualified individuals who took the UFO problem seriously, so that in the event of a Congressional hearing, there would exist an organization to which the Congress could turn."[8]

According to Saunders's account, "Hynek had had a cold and excused himself from our meeting early. After he left, McDonald brought up Low's controversial 'trick' memo. We were surprised that he knew anything about it. McDonald said that [NICAP's] Keyhoe had told him about it." Saunders went on to explain that he "had allowed Keyhoe to copy the memo on the day before Thanksgiving and had encouraged him to share the memo with members of NICAP's Board of Directors."

There is reason to question the foregoing Saunders account on at least one score—that he, Levine, and Armstrong were surprised that McDonald knew anything about the Low memo. Recall McCarthy's thesis statement that, at the time the Low memo was discovered, McDonald was covertly communicating with Armstrong and others in the Colorado project.

On December 28, 1967, barely two weeks after the secret meeting in Denver, McDonald wrote a leading French UFOlogist saying that he was "disappointed and disillusioned with Condon." McDonald added that "some confrontation is going to have to be effected. This is difficult to engineer. A number of us are working on that problem."[9] If McDonald wanted a confrontation, he need only have flown to Boulder and met with Condon and Low. What McDonald really wanted was to discredit the Colorado effort.

The coming months, as McCarthy wrote in his thesis, brought "the confrontation which McDonald desired. He played the major role in this episode, which turned on the infamous 'Low memo.' "[10]

On January 31, 1968, McDonald wrote a long letter to Low in which he criticized the conduct of the Colorado investigation. In his letter McDonald revealed that he had seen a copy of the Low memo and quoted portions of it verbatim. Interestingly, McDonald did not claim that he had been shocked by the contents of Low's memo, but said he was "rather puzzled by the viewpoints expressed there."[11]

McDonald's letter was buried in a stack of mail waiting for Low on his return trip. Mrs. Armstrong finally brought it to his attention on February 6. The next day, when Condon met with Saunders and Levine and they admitted having given a copy of the memo to McDonald, Condon fired both men. Shortly afterward Mrs. Armstrong resigned.

On February 9, McDonald sent a copy of the Low memo to the president of the National Academy of Sciences, which had agreed to review the Colorado study's final report. Two days later, McDonald also sent a copy of the memo to James Hughes, Office of Naval Research. Hughes was the contract monitor for several atmospheric research contracts the Navy had awarded to McDonald, and McDonald had covertly used these research funds for his UFO investigations and travels, with Hughes's tacit approval.

McDonald explained that he wanted to keep Hughes informed on the UFO scene in expectation of the time when he would formally seek Navy funds for UFO research, rather than bootlegging these activities. Not surprisingly, McDonald warned Hughes not to discuss such "explosive material" in letters sent to his office and that any discussion of these matters should be sent to McDonald's home.[12]

On April 30, 1968, coincident with the publication of Fuller's article in the May 14 issue of *Look* magazine, NICAP held a press conference in Washington to denounce the Low memo and the Colorado investigation.

During the question-and-answer period, I asked NICAP director Keyhoe, "What would you do if one of your employees were to go through your files and take certain papers out and send them to me, or to the Air Force, without your knowledge?" Keyhoe replied: "I'd probably fire him. I'd take a dim view of the disloyalty, not the papers."

When Keyhoe sharply criticized Condon for failing to make field investigations of UFO cases, I asked Keyhoe: "How many field investigations have you made in the past year?" Keyhoe replied: "About five and I made a great many by telephone." When I asked him to identify the five cases he investigated, Keyhoe replied: "Three of them I don't care to mention because one of them, one group consists of sightings the Air Force did not report to Condon." I then responded: "Well, could you mention the [other] two you did investigate?" Keyhoe declined to answer.

On the very same day that *Look* hit the stands and NICAP held its press conference, Congressman J. Edward Roush (D-Ind.) took to the floor of the House to denounce the Colorado investigation. Roush said: "There is a strong indication that the Colorado project will be known as the $500,000 fiasco. At the very least, grave doubts have arisen as to the scientific profundity and objectivity of the Colorado project." Roush urged that the USAF be relieved of its responsibility for investigating UFOs and that the job be turned over to the Congress!

Thanks to McCarthy's thesis, we now know that Roush's speech on the floor was part of McDonald's cleverly orchestrated plan. More than a year before, only several months after the Colorado investigation had begun, McDonald contacted Congressman Roush when he visited Tucson. And on March 3, 1967, McDonald wrote to Roush to "push for Congressional hearings" on UFOs.[13] McDonald continued to write Roush to urge him to hold these hearings. But, in late 1967, Roush replied that, while he too favored congressional hearings on UFOs, he thought they should be deferred until the Colorado study was completed.

Three months after the *Look* article was published and after Roush had denounced the Colorado effort, he arranged for the House Science and Astronautics Committee (of which he was a member) to hold a one-day "UFO symposium." It was characterized as a symposium rather than a hearing for good reason. Five of the six scientists invited to testify, including McDonald and Hynek, were strongly pro-UFO. The sixth, Carl Sagan, at the time was mildly pro-UFO. (Sagan later became a UFO-skeptic.)

Thanks to McCarthy's thesis, we now know that Roush allowed McDonald to select the six scientists invited to testify. McDonald used Roush's office as his base of operations and was authorized to use the congressman's telephone credit card for long-distance calls made outside

Roush's office. Roush's very one-sided UFO symposium provided grist for the mill of author John G. Fuller, who promptly produced a paperback book entitled *Aliens in the Skies: The New UFO Battle of the Scientists,* consisting largely of pro-UFO testimony presented at Roush's symposium. The book provided still another opportunity to browbeat Condon, Low, and the University of Colorado.

In November 1968, Congressman Roush was defeated at the polls. Shortly thereafter he was named to NICAP's board of directors.

The *Look* article, the NICAP press conference, and Congressman Roush's denunciation of the Colorado investigation, together with the publication of the book by Saunders and Harkins, occurred many months before the final report on the Colorado effort was made public, in early 1969.

In one of the several sections of the final report authored by Condon himself, he said he had not been aware of the controversial Low memo until it burst into public view. He noted that Low's memo represented "at most, preliminary 'thinking out loud' about the proposed project by an individual having no authority to make formal decisions."[14]

More important, Condon pointed out that one of Low's key recommendations in his memo, that the investigation focus on the psychology of people who report UFOs rather than investigate the phenomenon itself, was "exactly contrary to the procedure actually followed by the project." Condon added: "It should be evident to anyone perusing this final report, that the emphasis was placed where, in my judgment, it belonged: on the investigation of physical phenomena, rather than psychological or sociological matters."[15]

Condon could have added—but probably was too embarrassed to do so—that, if he and Low had wanted to conspire to create false impressions, then Condon would never have given his several public speeches expressing his strong skepticism about UFOs.

During the two years following publication of the Condon Report, as it is now known, until McDonald's tragic death, McDonald continued to give pro-UFO lectures to scientific groups around the country. I heard, or obtained copies of, many of those talks. *Never once did McDonald raise the issue of the "trick" memo in his talks before scientific groups.* Apparently he realized that Low's memo had served its purpose, to blacken the names of Condon, Low, and the University of Colorado in the public eye, and would carry no weight-of-argument among scientists.

Instead, McDonald chose to criticize the Colorado effort on the grounds that there were a number of UFO cases for which the team was not able to find rational, prosaic explanations. The same criticism is voiced today. Yet the fact that there were unexplained UFO cases demonstrates that the Colorado project did not avoid tackling challenging incidents. And the fact that these were published in the Colorado report shows that there was no attempt at coverup or censorship. In other words, contrary to the impression conveyed to large segments of the public by McDonald's well-orchestrated campaign, Condon and Low did not resort to skullduggery.

But what about those unexplained cases? One McDonald found especially impressive involved the crew of a USAF RB–47 electronic reconnaissance aircraft on a night training mission over the Gulf states. The Colorado team was handicapped in investigating this case because, when the pilot brought the case to the attention of Colorado investigators, he provided an incorrect date and the Colorado investigators were not able to locate original files and reports.

Later, when McDonald managed to locate the original files, I myself tackled this case and found it one of the most challenging I had ever undertaken. It required many hundreds of hours of effort, including locating twenty-year-old data on the antenna radiation pattern of an old radar and the schematic diagram for a decade-old electronic intelligence system. After a lot of work, and with a lot of luck, I was able to develop a prosaic explanation that was endorsed both by the RB–47 pilot and by the electronic intelligence (Elint) operator whose equipment was involved.[16]

Another decade-old UFO incident that had occurred in England also went unexplained in the Condon Report. My own lengthy investigation developed a prosaic explanation.[17] I suspect that the young scientist assigned to this case overlooked potential prosaic explanations because he was too eager to believe that some UFOs were extraterrestrial spaceships.

Still another of the unexplained cases in the Condon Report involved two photos of a UFO that resembled an inverted pie tin. The young scientist who investigated this case, and who has since become a well-known planetary scientist, was simply too credulous in accepting statements from seemingly honest farm folk. Later, this scientist revised his original view and concluded the photos were a hoax after considering evidence developed by Robert Sheaffer and my analysis.

All three of the unexplained cases in the Condon Report that I investigated proved to have prosaic explanations. I am confident that the others do also.

There are important lessons to be learned from the University of Colorado project. They are especially important if the U.S. government should ever decide to fund a scientific investigation into other claims of the paranormal, such as parapsychology.

If the government did decide to fund an investigation into parapsychology, for instance, who could be expected to volunteer for the effort? The most eager volunteers would be those scientists who already are investigating psi. Most, if not all, of them believe that psi exists or they would not be devoting their careers to the subject. But clearly this disqualifies them as impartial researchers.

What *experienced* scientist would volunteer to even temporarily abandon his or her present career in a field that seems promising for one that seems to be a pseudoscientific dead end? The bulk of the volunteers for such a project would be young, inexperienced scientists who would hope to become pioneers in a new field with profound implications. These young researchers would be hopeful of discovering evidence early in their careers that could make them famous. And it is this attitude that could make them vulnerable to becoming victims of self-delusion and ambition.

In retrospect, it is clear to me that at least some, if not many, of the young researchers who volunteered for the University of Colorado UFO investigation had what might be called "UFO-stars-in-their-eyes" hopes and ambitions. There is no hard evidence to show that they were as strongly pro-UFO as Saunders and Levine, but considerable circumstantial evidence to suggest that at least several of the young scientists were not as skeptical as they should have been.

Another criticism I would level at the Colorado project is that it undertook to investigate too many old UFO cases. The older the UFO incident, the more difficult it is to obtain the necessary hard data and the more flawed will be the recollections of those who were involved. Yet, if Condon and Low had opted to focus on current UFO cases, their critics would have accused them of ignoring the most impressive data.

If there had been UFO reports *only* during the late 1940s and the 1950s and none thereafter, the Colorado investigators would have had no choice but to investigate the old incidents. But UFOs were a continuing, ongoing phenomenon, with many hundreds of recent cases and many dozens that occurred while the Colorado investigation was in progress. These current cases could be considered a representative sample of the phenomena, and the Colorado should have focused their principle energies on these incidents. But I acknowledge that this is the wisdom of hindsight.

In summary: Under the difficult circumstances, Condon and Low probably did as good a job as was possible, especially considering that one of the three principal investigators already was at least a quasi-believer before the effort got under way.

One thing is now certain, however. The late Edward Condon, Robert Low, and the good name of the University of Colorado all were victims of a well-orchestrated plot to discredit them—a plot whose outlines would still be unknown but for the tragic death of its principal architect and the efforts of an enterprising Ph.D. candidate.

Condon's own brief comments on the skullduggery that occurred, contained in the final report, are illuminating: "I had some awareness of the passionate controversy that swirled around the subject, contributing added difficulty to the task of making a dispassionate study. . . . Had I known of the extent of the emotional commitment of the UFO believers and the extremes of conduct to which their faith can lead them, I certainly would never have undertaken the study."[19]

NOTES

1. David R. Saunders and R. Roger Harkins, *UFOs? Yes!—Where the Condon Committee Went Wrong* (New York: Signet Books), p. 81.
2. Ibid., p. 92.
3. Ibid., p. 139.
4. Ibid., p. 138.
5. Paul McCarthy, "Politicking and Paradigm Shifting: James E. McDonald and the UFO Case Study," Ph.D. diss. (University of Hawaii, 1975), p. 137.
6. Saunders and Harkins, *UFOs? Yes!* p. 134.
7. McCarthy, "Politicking and Paradigm Shifting," p. 170a.
8. Saunders and Harkins, *UFOs? Yes!* p. 179.
9. McCarthy, "Politicking and Paradigm Shifting," p. 142.
10. Saunders and Harkins, *UFOs? Yes!* p. 249.
11. Ibid., p. 249.
12. McCarthy, "Politicking and Paradigm Shifting," p. 145.
13. Ibid., p. 177.
14. *Final Report of the Scientific Study of Unidentified Flying Objects* (New York: Bantam), p. 549.
15. Ibid.
16. Philip J. Klass, *UFOs Explained* (New York: Random House), chapters 19 and 20.
17. Ibid., chapter 21.
18. Ibid., chapter 15.
19. *Final Report of the Scientific Study of Unidentified Flying Objects,* p. 548.

3.
PSYCHOLOGY AND UFOS
Armando Simón

Investigations of UFOs have for the most part been performed either by nontechnical personnel or by professionals whose realms of expertise extend only into the natural sciences, in particular, astrophysics. It is only logical that astronomers would be involved, since flying saucers from the very beginning were assumed to be extraterrestrial spacecraft (Menzel and Boyd 1963; Vallee 1965). However, mundane meteorological explanations for sensational cases of UFOs (translate: blobs of light) traveling at an apparent 200 mph and making 90° turns were emotionally unsatisfactory. Even when the investigator patiently explained that any living organism inside such a "craft" would have become grape jelly by a 90° turn, due to the force of inertia, such explanations were waved aside with the frustrating response that "the laws of science as we now know them may not apply to the craft" or that the assumed aliens were far advanced technologically and could have overcome such a petty obstacle.

Refusal to believe in "down to earth" solutions resided in several types of advocates (Jacobs 1975). One was the cynical type, who would financially milk the phenomenon for all it was worth. Another was the "contactee" (a person who claimed to have been picked up by angelic-looking aliens who were Christians, given a ride to Mars or Venus, maybe even meeting Jesus, and given a mission to spread pacifism on Earth) and his or her followers (Menger 1959). A third type of advocate would adhere to any belief even remotely connected with the occult. Last, there

This article originally appeared in the *Skeptical Inquirer* 8 (Summer 1984). Reprinted with permission.

were groups of people who sincerely believed that the UFOs they saw were extraterrestrial ships, such beliefs being reinforced by the sometimes spectacular effects present in the sightings. Yet none of the believers ever questioned their basic assumption: Why were the flying saucers automatically interpreted as being from outer space and piloted by aliens, inasmuch as they were a novel phenomenon with no previous clue as to their origin or occupants? To this day, this question has not been considered by either the UFO skeptics or the believers.

Psychologists studying the UFO phenomenon were scarce until after the conclusion of the University of Colorado Condon Report in 1968. The lateness of the involvement of psychologists is both surprising and unexplainable, since early on there was much speculation voiced about the psychological components in sightings of flying saucers. Most of the earlier questions and opinions alluding to psychological ramifications had been rather crude: "Are people who see flying saucers crazy?" "Flying saucers go away when the silly season ends." "One cannot doubt the reports made by reliable professionals like policemen, doctors, and pilots." "Did the people who saw the UFOs last night really imagine everything?" (Jacobs 1975; Hynek 1974).

Even so, the phenomenon strongly implied a role for psychologists. When a few did finally get involved, they faced the same problems natural scientists had faced, namely, that a UFO could not be seized for study, nor would it stay in one place long enough for instrument readings, much less a research design. Worse, they had to shake off the sensationalism promoted by the mass media.

The research presented here falls within various branches of psychology and is so grouped. No new specialty within psychology has emerged to investigate this phenomenon. The degree of interest within a particular specialty has been sporadic and subject to a lack of systematization. Personality researchers, for example, by all expectations should have taken the personality correlates of skeptics, contactees, hoaxers, and reporters and nonreporters of sightings in order to compare the profiles. Yet few, if any, psychologists have done so. In contrast, the field of attitude-formation has enjoyed comparatively heavy investigation vis-à-vis UFOs, as will be seen.

PSYCHOANALYSIS

In spite of their sexually suggestive elliptical and cigar-shaped descriptions, UFOs have drawn relatively little attention from the psychoanalytic field. Wilhelm Reich, toward the end of his turbulent career and before his research papers were burned by the Food and Drug Administration, saw a UFO and was of the opinion that it was of extraterrestrial origin and warlike in its intentions (Boadella 1973). His interpretations were not psychological, nor was the research he engaged in during the latter part of his life—such as his work on "orgone energy," which he suggested could be put to good use against flying saucers in the forthcoming war of the worlds. Ironically, Reich, like another famous psychoanalyst, Sigmund Freud, became somewhat mentally unstable in his later years.

It should not be too surprising that a psychoanalyst as multifaceted in his interests as Carl Jung (1961) would turn his attention to flying saucers. He was one of the few to do so with more than a passing interest. Jung first gave his views on the subject in a 1954 Swiss weekly, and these views were subsequently distorted by the world press (Jung 1959). When he issued a statement correcting the sensationalist version, he was curiously ignored by the press, a phenomenon similarly noted by UFOlogists Philip Klass and J. Allen Hynek. "The moral of this story," Jung wrote in his book *Flying Saucers,* "is rather interesting. As the behavior of the press is a sort of Gallup test with reference to world opinion, one must draw the conclusion that news affirming the existence of UFOs is welcome, but that skepticism seems to be undesirable. To believe that UFOs are real suits the general opinion, whereas disbelief is discouraged. This creates the impression that there is a tendency all over the world to believe in saucers and to want them to be real." Reviewing the pertinent case histories, including his patients' art works, and extensively reading the UFO literature, he suggested a hypothesis. He started with the fact that the mandala has at all times, in all cultures, been the symbol of order, of wholesomeness, of a desire to bring sanity out of insanity. Further, he affirmed that UFOs were visionary rumors and that in order to have a visionary rumor the following are prerequisites: (1) unusual emotion, (2) emotional tension, and (3) projection.

Essentially, Jung stated that, in the 1950s and 1960s, mankind was faced with the dilemma of nuclear annihilation, on the one hand, or totalitarian subjugation due to Soviet imperialism, on the other. (Needless to say, his works were banned in the USSR.) Therefore, flying saucers were projections from the unconscious that, because of their mandala shapes and their alleged heavenly origins and technological superiority, had an air of salvation about them. Because of our modern rationalism, we interpreted such visions of longing and deliverance not as heavenly, other-worldly angels, but as mechanical, other-worldly spaceships. *Such projections, incidentally, would come from quite normal persons.* It should be further noted that the peak years for flying saucer sightings in the United States (1952, 1957, 1966) were years when the world dilemma referred to above was most threatening to the American public (the Korean War and presidential elections, the highly traumatic Sputnik, and the Vietnam War).

One other psychoanalyst became interested in UFOs. Berthold Schwarz (1968, 1969), an American psychiatrist, began a decade later to inquire about a totally different aspect of the phenomenon, to wit, the frequency of UFO themes in the narratives of psychiatric patients as well as the question of sanity in relation to sightings of UFOs. In a local journal of medicine in New Jersey, Schwarz reported that, of the thousands of patients he had seen and of the thousands of patients in a nearby county mental hospital, not one had ever reported any UFO experiences. Conversely, all of the sighters that Schwarz interviewed seemed to be quite stable and without any symptomatology. The only case that I myself have run across in the literature has been of a case study in Arthur Janov's book on primal therapy, wherein flying saucers did play an important role in the patient's delusions.

HYPNOSIS

The use of hypnosis, more than any other aspect of psychology in UFO research, has gained much publicity. Not only have popular magazines like *Omni, Psychology Today,* and *Astronomy* treated the topic, but a TV dramatization of the well-known Betty and Barney Hill abduction—which some psychoanalysts have interpreted simply as a case of a *folie à deux*—was aired in 1975, nearly ten years after the original incident. There seems to be a set pattern in nearly all abduction cases. This pattern, and the fact that most of the "time loss" cases occurred after the 1966 publication of *The Interrupted Journey* (Fuller 1966), implies mass suggestion.

An abduction case usually goes this way: A person is in a deserted area of the country, possibly with some relatives or friends, sees a UFO, and after some time is trapped by aliens who physically examine and then return the individual, sometimes with amnesia (Sprinkle 1979; Fuller 1966; Wells 1981).

Some doubts have begun to be cast upon the validity of narratives given under hypnosis. Alvin H. Lawson (1977) found 16 people who were not well acquainted with UFO lore. He had a clinical hypnotist hypnotize them into believing they were inside a flying saucer and then had them describe their "experience." Their accounts were quite similar in detail to the actual reports of abductions and included medical examinations. His conclusions were presented at the American Psychological Association in 1976.

An experimental objection to the Lawson study was carried out by Allen Hendry (1979), the chief investigator at the Center for UFO Studies. A woman had seen a UFO in 1978 and claimed to have been in telepathic communication with the aliens, who did not land; she also experienced nausea. Upon obtaining additional information, Hendry found a conventional explanation for the object seen but did not reveal the identity of the object to the sighter (the woman, incidentally, had claimed never to have believed that anything like the story above could be true). Hendry then hired a hypnotherapist. Under hypnosis, the woman did *not* invent any additional details, and her account was in the past tense, unlike that of true "abductees." In other words, under hypnosis, she stated she had seen a UFO and not that she had been abducted, nor that she had been medically examined by the aliens. One last point: Her Minnesota Multiphasic Personality Index profile was evaluated as not coming from a psychologically healthy person.

It is surprising, however, that to date few have looked at the abduction cases in light of the controversy enveloping hypnosis as a whole. The controversy revolves on the question, What exactly *is* hypnosis? There are two theories. One interprets hypnosis in its traditional sense, that of inducing a trance state as a gateway to the unconscious. The alternative interpretation is that the "hypnotizing" of a subject leads to a role-playing state, the role being that of a "trance state"; the degree of influence over a subject (even to the point of causing warts and ignoring pain) is directly related to how suggestible that person is (Spanos and Barber 1974). This second theory is somewhat related to the objection made over the abduction cases, that the hypnotherapist is subtly influencing the session. Thomas Szasz has given a relevant definition in *The Second Sin*: "Hypnosis: two people lying to each other, each pretending to believe both his own and his partner's lies."

ATTITUDES

Attitude formation is one of the most well-traveled research paths in both psychology and sociology. Nearly every topic imaginable that one could have an opinion on has been investigated and its correlates explored. Most of the research on attitudes toward UFOs/flying saucers/extraterrestrial-life was done during and after publication of the Condon Report. Indeed, for many reasons the 1966 wave of sightings proved to be the watershed for the beginning of all reputable research on the phenomenon (Simón 1979a). The attitude studies fell in two general directions: one was to identify the various outlooks toward flying saucers, their interrelationships, and who held such beliefs; the other was to use the belief in such concepts to test a particular theory. An example of the former is the periodic Gallup Poll, which has consistently seen a rise in the percentage of the population admitting to having seen a UFO and to believing there is life in the universe beyond our planet. The Gallup data, incidentally, have been used for further secondary analysis by other researchers.

Aldora Lee (1969) emphasized the importance of studying attitudes toward flying saucers by pointing out that almost always there is no physical evidence, just verbal reports, and that the reports are colored by the beliefs held by the witnesses. Examining the Gallup data and conducting polls on her own for the much-maligned Condon project, Lee found that (a) age, education, and geographical region were related to whether respondents believed in extraterrestrial life, (b) sighters do not differ from nonsighters with respect to sex, age, education, or geographical region, (c) adults, compared with teenagers, tend to be negatively inclined toward the possibility of UFOs' extraterrestrial origin.

In establishing the various attitudes toward UFOs independently, Simón (1979b) and Saunders (1968) have employed very complicated statistical techniques in order to ascertain the very obvious. For example, a sixty-three-item questionnaire that Simón (1979b) administered to college students and members of the National Investigations Committee of Aerial Phenomena (NICAP), a UFO group, upon being factor-analyzed, brought out ten obvious attitude factors: belief in extraterrestrial visitation, skepticism of UFOs, belief in the occult, extrapolation of nature, psychological explanations, disbelief in contactees, witnessing, belief in extraterrestrial life, reluctance to report a sighting, belief in possibilities.

On a different plane altogether is Lawrence Littig's (1971) brief but important study. Littig was interested in matching the belief in life in other worlds with degree of *affiliation motivation* (the desire to be accepted and loved by others). Giving his subjects a test for affiliation motivation and separating the high and low scorers, he compared their responses to the statement, "Further research on UFOs (Unidentified Flying Objects, e.g., 'flying saucers') will reveal that they come from outer space." He found that subjects scoring strongly in affiliation motivation tended to significantly agree with the statement. Littig furthermore stated that a peopled universe is more reassuring than an empty one and that there are psychological processes involved in the belief in extraterrestrial life, but that such processes tended to be normal ones rather than abnormal and irrational.

Littig's study was indirectly supported by P. Fox (1979). Asking her subjects what they thought UFOs were, she matched her answers with attitudinal and background data. She found that the best predictor of an extraterrestrial interpretation of UFOs was agreement with the statement that there are intelligent life forms on other planets.*

Littig's study was unfortunately overlooked within the field of UFOlogy during the controversy in *Science* sparked by Warren (1970), possibly because *Science,* unlike a specialized journal, is read by all disciplines. Warren examined the 1966 Gallup Poll data in view of the *status inconsistency theory.* Essentially, this theory states that individuals who do not have equivalent status in ethnicity, education, income, or intelligence are bound to feel that they are "outsiders" and hence are under tension. This tension will find expression in asocial actions. Warren came to the conclusion that sighters of flying saucers upheld the theory. He maintained that upon breakdown of the collected data, the Gallup Poll verified that status inconsistents were sixteen times more likely to see UFOs. In a UFO newsletter rebuttal, R. Leo Sprinkle of the University of Wyoming countered that the real question was the difference between sighters who report the UFOs seen and sighters who do not report their sightings. Hynek (1974), in turn, dismissed the hypothesis in his book *The UFO Experience* because in one particular sighting the witnesses did not appear to be status inconsistent. The best critique of the Warren study, however, came from Saunders, who delivered a report at the American Institute of Aeronautics and Astronautics in 1975. Saunders simply pointed out that Warren's paper was devoid of any statistical analysis. His own analysis of other, independent data seemed to indicate that it was status *consistency* that was predictive of sightings, a finding supported by other studies (Westrum 1979; Fox 1979). In my view, this makes more sense, since UFOs are almost universally accepted in North America. To date, no more attempts at settling the controversy have surfaced anywhere.

PERCEPTION

Logically, perceptual psychologists should have been very productive on the subject of UFOs. Yet with one important exception (Haines 1980), the extent of research on UFOs' (mis)perception has been to cite case studies and a few experiments, some of them involving the common *autokinesis* and *autostasis* effects (Wertheimer 1968). For example, Carl Sagan and Thornton Page (1972) in *UFOs—A Scientific Debate* cite a case study

---

*Both of these studies at least explain why it is that nearly every time I reply in the negative when asked if I believe flying saucers are visits from space, I am immediately asked the irrelevant, nonsequitur question, "But don't you believe in life on other worlds?" The same thing, I am sure, happens to others.

wherein several observers witnessed the reentry of Zond IV and interpreted it as a UFO. Reports came in of its having windows, starting grass fires, frightening a dog, and causing an observer to be unexplainably overcome with sleep. Even the Condon Report, which was specifically charged with the study of the psychological components of UFOs, limited its treatment of perception to a couple of general textbook chapters in its ballast section; there was only one true experimental study in the report, that of Aldora Lee cited above.

In a series of tests, Haines (1979, 1980) asked two groups of people to draw a UFO, comparing the drawings by the group that reported having previously seen a UFO with those of the one that had not seen a UFO. Both tended to give the basic dimensional characteristics, suggesting that there is an underlying consensus of what a UFO should look like. Nevertheless, the former group tended to draw fewer openings, apertures, insignia, domes, and landing gear than the latter group. Haines then investigated the reliability of potential drawings of UFOs by presenting a visual stimulus to a group of subjects and asking them to draw it, the stimulus being allowed to be viewed as long as desired. Errors were made. Even bigger errors were made during a UFO investigator workshop, where participants were asked to draw the UFO that was verbally described by a witness; the investigators' drawings compared poorly with the original witness's. A large amount of "investigative experience" in terms of years did not necessarily guarantee greater accuracy. Haines ended his series of experiments by suggesting that further research was indicated, a classic of understatement.

Addressing himself to this same problem, but more with the aim of aiding future investigators, Roger Shepard (1979) noted that test results based on recognition are more valid than tests using verbal recall. He compiled an exhaustive array of UFO drawings based on photographs, as well as computer-generated silhouettes of UFOs and a list of terms used in the past in order to describe observed phenomena in future sightings.

Paradoxically enough, despite the involvement of psychologists in investigating UFOs, it is a nonpsychologist who has come forth with the most testable hypotheses amenable to perceptual research. Philip J. Klass (1974), who incidentally has urged involvement by psychologists in the UFO arena, presented several propositions in his book *UFOs Explained*:

1. Basically honest and intelligent persons who are suddenly exposed to a brief, unexpected event, especially one that involves an unfamiliar object, may be grossly inaccurate in trying to describe precisely what they have seen.
2. Despite the intrinsic limitations of human perception when

exposed to brief, unexpected and unusual events, some details recalled by the observer may be reasonably accurate. The problem facing the UFO investigator is to try to distinguish between those details that are accurate and those that are grossly inaccurate. This may be impossible until the true identity of the UFO can be determined, so that in some cases this poses an insoluble problem.

3. If a person observing an unusual or unfamiliar object concludes that it is probably a spaceship from another world, he can readily adduce that the object is reacting to his presence or actions when in reality there is absolutely no cause-effect relationship.

4. News media that give great prominence to a UFO report when it is first received, subsequently devote little if any space or time for reporting a prosaic explanation for the case when all the facts are uncovered.

5. No human observer, including experienced flight crews, can accurately estimate either the distance/altitude or the size of an unfamiliar object in the sky unless it is in very close proximity to a familiar object whose size or altitude is known.

6. Once news media coverage leads the public to believe that UFOs may be in the vicinity, there are numerous natural and man-made objects which, especially when seen at night, can take on unusual characteristics in the minds of hopeful viewers. Their UFO reports in turn add to the mass excitement which encourages still more observers to watch for UFOs. This situation feeds upon itself until such time as the news media lose interest in the subject, and then the "flap" quickly runs out of steam.

7. Whenever a light is sighted in the night sky that is believed to be a UFO and this is reported to a radar operator, who is asked to search his scope for an unknown target, almost invariably an "unknown" target will be found. Conversely, if an unusual target is spotted on the radarscope at night that is suspected of being a UFO, and an observer is dispatched or asked to search for a light in the night sky, almost invariably a visual sighting will be made.

SOCIAL PSYCHOLOGY

The studies that fall into the social psychology category are somewhat similar to those under the section on perception. There is a lack of reported investigation in proportion to the wealth of potential experiments that could be conceived and concluded.

By coincidence, however, a classic study in social psychology involved flying saucers (Festinger, Riecken, and Schachter 1956). *When Prophecy Fails* detailed the 1954 infiltration of a contactee group for the purpose of testing a theory of *cognitive dissonance.* A cult group predicted the rescue of its members by flying saucers following the destruction of the world. Needless to say, the predictions did not bear out. Whereas before the doomsday date arrived the cult leaders were highly secretive and skeptical of both the news media and new converts, the reverse came to be the case after disconfirmation of the rescue prophecy. Not only did publicity and proselytizing begin, but, paradoxically, their faith was reinforced, especially if they had made deep irreversible commitments of a social and financial nature.

A study by H. Buckner (1966) of Concordia University in Montreal gave some rough parameters of a California contactee group. The members had an average age of sixty-five, were widowed or single women, had poor health, were interested in the occult, and had little formal education. The men either were young schizophrenics or were aged with advanced senility. Criticism of doctrines was discouraged; all ideas were legitimate.*

Another area of research within social psychology has been the influence of the mass media on the rate of UFO sightings. Simón (1979a) unsuccessfully attempted to answer this question by correlating the number of sightings per year with the number of science fiction movies per year. It was found that such a correlation would be spurious due to a number of idiosyncrasies within the data: (1) the making of a film took a long time, sometimes over a year from its inception, (2) there was a *qualitative* difference between films, whose appeal to the public was elusive to statistical analysis, and (3) the year of initial release of a film was no indication of that film's duration in theaters, nor of its return at later times. Basically, the problems encountered were typically those found within unobtrusive data (Campbell and Stanley 1963; Simón 1979c). Nevertheless, the increase of such films after the first official sighting of a flying saucer, in 1947, was suggestive, as was the thematic material found in the films themselves.

A second attempt at gauging the effect of the mass media on UFO sightings again brought mixed results (Simón 1981). Again, this was partly due to the statistical analyses. Statistical tests were run and significant results found for all the numerical data presented, but it was pointed out by the editors, quite correctly, that the underlying assumptions for the tests were inapplicable to the data, and unnecessary due to the dramatic changes evident in the data.

This second study attempted to (1) objectify public interest in fads

---

*From personal experience with a similar Kansas-based group, I can verify these observations. One thing that neither of the investigators mentioned is the fact that the contactees' lectures and readings are very bland and boring, with nothing of the sensationalism found in noncontactees' reports.

through numerical tabulations of magazine and newspaper articles, films and television shows, and the number of public library books checked out by readers on the particular topic of interest, (2) show that the development of a buildup of UFO sightings follows the same pattern as other fads, with a strong implication that the mass media are a causal factor.

The underlying assumption to both studies is that the number of unidentified flying objects remains basically stable but that the public's interest in them fluctuates due to particular events in the culture (individual countries tend to have sporadic increases in reported sightings).

CONCLUSIONS

It is apparent that some of the above-mentioned studies suffer from *experimenter bias,* possibly due to the researcher's interest that UFOs be acknowledged in their true identity, either as extraterrestrial spacecraft or as misperceived terrestrial phenomena. Any psychologist planning on doing any long-term research involving UFO sightings or UFO sighters would do well to stay clear of the seductive yet vacuous UFO subculture in order to maintain the integrity of his research. To be sure, the investigation of certain areas (such as the contactees) necessitates delving deeply into the subculture.

In addition, there is a distasteful impression that instead of research data and theories being worked out and applied to society, an apparently ludicrous and disreputable aspect of society is being thrust onto a serious subject, be it psychology, astronomy, or physics.

It is frustrating to summarize what research has been done. Part of the problem is the aimlessness of the literature as a whole. Little relevancy connects one study with another except when one author pursues a question throughout several studies. Nevertheless, in light of the studies mentioned above, several conclusions can be affirmed:

1. There is a strong indication that those UFOs whose existence can be objectively verified, and are usually blobs of light in the sky, are the physical equivalent of Rorschach inkblots. They will be reported according to the reporter's mental set.

2. Belief in flying saucers and extraterrestrial life is positively correlated with youth, high income, increased education, and being male.

3. To believe that UFOs are piloted by "little green men" is the cultural norm in North America, and not, as is put forth, a minority viewpoint. It is not a sign of mental illness.

4. There is a strong indication that the rate of flying-saucer sightings fluctuates according to the amount, as well as the quality, of coverage in the mass media.

5. The "contactee" cult groups, mostly based in California, are composed primarily of senile or schizoid members. They tend to be religious and low-key.

6. Due to (3) and (4), there appears to be a cultural consensus of what a "UFO" is supposed to look like and what the "aliens" are supposed to look like.

Numerous questions remain unanswered in each specialty within psychology. Some areas, such as personality correlates, remain virtually untouched and offer a fine repository of potential data for theses or dissertations. If properly exploited, the whole topic of UFOs/flying-saucers can yield much useful information that can be applied to many areas other than the UFO subculture. Only the surface has been scratched. Future research must incorporate the experimental rigors found elsewhere in the field of psychology.

REFERENCES

Boadella, D. 1973. *Wilhelm Reich: The Evolution of His Work.* New York: Dell.

Buckner, H. 1966. "Flying Saucers Are for People." *Trans-Action* 3: 10–13.

Campbell, D. T., and J. C. Stanley. 1963. *Experimental and Quasi-Experimental Designs for Research.* Chicago: Rand McNally.

Festinger, L., H. Riecken, and S. Schachter. 1956. *When Prophecy Fails.* Minneapolis: University of Minnesota Press.

Fox, P. 1979. "Social and Cultural Factors Influencing Beliefs about UFOs." In *UFO Phenomena and the Behavioral Scientist,* edited by R. Haines, 20–42. Metuchen: Scarecrow Press.

Fuller, J. 1966. *The Interrupted Journey.* New York: Dial.

Haines, R. F. 1979. "What Do UFO Drawings by Alleged Eyewitnesses and Noneyewitnesses Have in Common?" In *UFO Phenomena and the Behavioral Scientist,* edited by R. Haines, 358–95. Metuchen: Scarecrow Press.

———. 1980. *Observing UFOs.* Chicago: Nelson-Hall.

Hendry, A. 1979. "Trance Figures." *Omni* 1: 32–33.

Hynek, J. 1974. *The UFO Experience.* New York: Ballantine Books.

Jacobs, D. 1975. *The UFO Controversy in America.* Bloomington: Indiana University Press.

Jung, C. 1959. "A Visionary Rumor." *Journal of Analytical Psychology* 4: 5–19.

———. 1961. *Flying Saucers.* New York: Signet.

Klass, P. 1974. *UFOs Explained.* New York: Vintage.

Lawson, A. H. 1977. "Hypnosis of Imaginary UFO 'Abductees.' " *Journal of UFO Studies* 1: 8–26.

Lee, A. 1969. "Public Attitudes Toward UFO Phenomena." In *Scientific Study of Unidentified Flying Objects,* edited by E. U. Condon, 209–43. New York: Bantam.

Littig, L. 1971. "Affiliation Motivation and Belief in Extraterrestrial UFOs." *Journal of Social Psychology* 83: 307–308.

Menger, H. 1959. *From Outer Space.* New York: Pyramid.

Menzel, D., and L. Boyd. 1963. *The World of Flying Saucers.* Garden City: Doubleday.
Sagan, C., and T. Page. 1972. *UFOs—A Scientific Debate.* New York: W. W. Norton.
Saunders, D. 1968. "Factor Analysis of UFO-related Attitudes." *Perceptual and Motor Skills* 27: 1207–18.
———. 1975. "Extrinsic Factors in UFO-reporting." American Institute of Aeronautics and Astronautics (AIAA), Thirteenth Aerospace Sciences Meeting, January 20–22.
Schwarz, B. 1968. "UFOs: Delusion or Dilemma." *Medical Times* 96: 967–81.
———. 1969. "UFOs in New Jersey." *Journal of the Medical Society of New Jersey* 66: 460–64.
Shepard, R. N. 1979. "Reconstruction of Witnesses' Experiences of Anomalous Phenomena." In *UFO Phenomena and the Behavioral Scientist,* edited by R. Haines, 188–224. Metuchen: Scarecrow Press.
Simón, A. 1979a. "The Zeitgeist of the UFO." In *UFO Phenomena and the Behavioral Scientist,* edited by R. Haines, 43–59. Metuchen: Scarecrow Press.
———. 1979b. "Systematic Replication of Saunders' (1968) Attitude Factors." *Perceptual and Motor Skills* 48: 1199–1210.
———. 1979c. "Some Shortcomings of the Unobtrusive Methodology in Studying the Effects of the Mass Media: A Personal Account." *Southern Journal of Education Research* 13: 161–74.
———. 1981. "A Quantitative, Nonreactive Study of Mass Behavior with Emphasis on the Cinema as Behavioral Catalyst." *Psychological Reports* 48: 775–85.
Spanos, N., and T. X. Barber. 1974. "Towards a Convergence in Hypnosis Research." *American Psychologist* 29: 500–11.
Sprinkle, R. L. 1979. "Investigation of the Alleged UFO Experience of Carl Higdon." In *UFO Phenomena and the Behavioral Scientist,* edited by R. Haines, 225–57. Metuchen: Scarecrow Press.
Vallee, J. 1965. *Anatomy of a Phenomenon.* New York: Ballantine.
Warren, D. 1970. "Status Inconsistency Theory and Flying Saucer Sightings." *Science* 170: 599–603.
Wells, J. 1981. "Profitable Nightmare of a Very Unreal Kind." *Skeptical Inquirer* 5, no. 4: 47–52.
Wertheimer, M. A. 1968. "Case of 'Autostasis' or Reverse Autokinesis." *Perceptual and Motor Skills* 26: 417–18.
Westrum, R. M. 1979. "Witnesses of UFOs and Other Anomalies." In *UFO Phenomena and the Behavioral Scientist,* edited by R. Haines, 89–112. Metuchen: Scarecrow Press.

4.
THE "TOP-SECRET UFO PAPERS" NSA WON'T RELEASE

Philip J. Klass

If you chance to catch Stanton T. Friedman, UFOlogy's most colorful spokesman, in one of his frequent television, radio, or lecture appearances, you will hear him accuse the U.S. government of a UFO coverup that he calls a "Cosmic Watergate." Friedman, a nuclear physicist turned UFO lecturer, is a P. T. Barnum-type showman who typically charges that the National Security Agency (NSA) is withholding "160 top-secret UFO documents."

As proof, Friedman holds up several pages of a heavily censored, once top-secret petition submitted by NSA to the U.S. District Court in Washington explaining why release of the documents would likely "damage . . . our national security." NSA's position was endorsed by a U.S. District Court, a subsequent three-judge Federal Court of Appeals, and the U.S. Supreme Court. Thus NSA *is* withholding 156 UFO-related papers—almost the 160 that Friedman typically claims.

This might seem to show that Friedman is correct and there is indeed a U.S.-government UFO coverup. But in reality, it is Friedman who is guilty of withholding information from the public about these NSA papers—information that would challenge his claims.

Friedman typically identifies NSA as the nation's largest and most secretive intelligence agency, which is true. But he never describes its several missions (which are known to him as a result of reading James Bamford's 1982 book about NSA, *The Puzzle Palace*) that could explain NSA's actions.

This article originally appeared in the *Skeptical Inquirer* 14 (Fall 1989). Reprinted with permission.

One of NSA's primary missions is to eavesdrop on radio communications of potentially hostile countries, referred to as "communications intelligence," or COMINT. A second mission is to "crack" the cryptographic codes of other countries in order to decipher their intercepted communications. NSA's third mission is to develop cryptographic techniques for U.S. government and military agencies that it hopes will be impervious to being cracked by other nations.

According to the NSA petition to the U.S. District Court, the 156 "records being withheld are COMINT reports that were produced between 1958 and 1979." This means that these are decoded transcripts of intercepted messages from foreign government sources, most likely Soviet-bloc countries. They may have been intercepted by U.S. agents in the USSR or in Soviet-bloc countries, or even from covert facilities in "neutral" countries.

If any of them contained "smoking gun"-type "UFO secrets"—for example, revelations that the USSR knew that UFOs were extraterrestrial craft—it would be foolish for the U.S. government to try to maintain a "UFO coverup," since it could be exposed at any time by Soviet leaders. The relatively small number of such records in NSA's possession, collected over a twenty-one-year period, indicate very scant interest in UFOs by Soviet-bloc countries. The average is less than one a month.

Friedman never acknowledges that NSA might have good reason, other than a UFO coverup, to withhold such documents. He must surely recognize that they could reveal sources and facilities unknown to the Soviet bloc and, more important, *which cryptographic codes have been cracked by NSA and are no longer secure.* While the USSR itself may no longer be using some of the older codes, when they are replaced with new ones they are typically given to and used by other countries in the Soviet bloc.

Fortuitously, Tom Deuley, a man with a strong interest in UFOs, went to work for NSA in mid-1978 and was employed there for four years. He was at NSA during the period when Citizens Against UFO Secrecy (CAUS) was trying to force the agency to release the documents. Deuley is now active in the UFO movement, serving as an official in two UFO organizations: the Mutual UFO Network (MUFON) and the Fund for UFO Research (FUFOR).

In late June 1987, Deuley presented a paper at a MUFON conference in Washington entitled "Four Years at NSA—NO UFOs." In his paper Deuley said he had transferred to NSA headquarters just prior to attending the 1978 MUFON conference in Dayton, Ohio. "Before making the trip," Deuley said, "I felt it was necessary to let NSA know I had an interest in UFOs. . . . Within a week I had an appointment with some administrative officials to discuss my trip to Dayton and my interest in UFOs."

Based on that meeting, Deuley said he did not "get any feeling that they [NSA] even cared about UFOs."

As a result of the meeting, Deuley said, he "met several other persons at NSA, and from other agencies, who had maintained their own interest in UFOs over the years," and these NSA associates sent him newspaper clippings and cartoons about UFO incidents.

Because of Deuley's interest in UFOs, he was one of those selected by NSA to review its UFO-related material. Said Deuley: "I believe I saw or held copies of the large majority of the documents [that were] withheld in that FOIA suit. Though there may have been exceptions among the documents I *did not* see, none of the documents I was aware of had any information of scientific value."

The former NSA employee told the MUFON conference audience: "I did not see any indication of official NSA interest in the subject [UFOs].

"I did not see any exchange of material indicating any form of follow- up activities. . . . I did not see any indication of real involvement other than the existence of the documents themselves."

Deuley endorsed NSA's withholding the material, noting the need to protect intelligence sources and methods. "It is clear to me that the possibility of damage to national security sources and methods far outweighs the value of the information under question."

Deuley concluded that if NSA was "involved with UFOs in any active way, I would have at least caught a hint of it in their treatment toward me or that with my openness about the subject, some informal contact would have mentioned it. Because neither of these occurred, *I concluded that UFOs do not have any importance at NSA*" (emphasis added).

Because I was attending another concurrent session at the MUFON conference and did not hear Deuley's paper, I wrote him to obtain a copy.

When he sent it, he wrote that his paper was aimed at "deterring UFO investigators from wasting their time trying to get at the papers that NSA was allowed to withhold." He added: "The documents . . . are not worth the effort, in terms of forwarding the effort of UFO research."

On July 25, 1987, I sent a copy of Deuley's paper to Stanton Friedman, wondering if he would accept the views of a first-hand observer who was a fellow "UFO-believer." Several months later, Friedman and I appeared on a television show in Portland, Oregon, and he again whipped out the heavily censored NSA court petition for the television audience as proof of a "government coverup." Friedman made no mention of the content of Deuley's MUFON paper.

More recently, on December 9, 1988, when Friedman and I participated in a talk show on Seattle radio station KING, he was asked to document his charge of a government UFO coverup. Friedman responded: "The National Security Agency admits it has 160 UFO documents. They're highly classified. It not only refuses to release them. . . ." Again, no mention of Deuley's statements.

The foregoing may provide a useful perspective when one chances to see or hear Friedman make his Cosmic Watergate charge and cite the NSA papers to support his claim.

5.
THE AVRO VZ–9 "FLYING SAUCER"
William B. Blake

The claims that the U.S. government recovered "crashed flying saucers" in 1947, together with the many "UFO" sightings since that date, raise an interesting question. Did the United States ever attempt to build and fly a manned flying saucer? The answer is yes. In the 1950s the U.S. funded Avro Aircraft Limited of Canada to build a saucer-shaped craft. What follows is a summary of that effort from historical records, including once "secret" photos of an experimental Avro saucer and other concepts for saucer-shaped craft. Because of serious, intrinsic stability problems, the U.S. Air Force/Army flying-saucer effort ended in failure a quarter-century ago.

In 1952, Avro Aircraft Ltd., located near Toronto, began a design study for a supersonic fighter-bomber airplane with a circular wing. The study was funded by a $400,000 contract from the Canadian government. The vehicle was intended to take off and land vertically, like a helicopter. The idea was to duct the jet exhaust to form a peripheral curtain of air underneath the vehicle. This would create a cushion of air on which the vehicle could "float." To transition from hover to high-speed flight, the jet exhaust would slowly be directed aft. Lift would be provided by the circular wing.

After the initial contract, the Canadian government abandoned the project as being too costly. Enough progress had been made, however, to interest the U.S. government. In July 1954, the first of two air force contracts totaling $1.9 million was awarded to Avro for further study. Avro

---

This article originally appeared in the *Skeptical Inquirer* 16 (Spring 1992). Reprinted with permission.

added $2.5 million of company funds to the effort and completed a series of design studies and small-scale tests on a vehicle designated the P.V. 704 (U.S. designation, System 606A). The 606A design was almost 30 ft. in diameter with a maximum weight of 27,000 lbs. and design speed over 1,000 mph.

The army became interested in the circular-wing concept and convinced the air force to redirect its effort in 1958. The army felt that the circular wing could fit in with its plans to develop a "flying jeep" for improved battlefield survivability. The air force agreed because it felt a small, subsonic research vehicle could be used to demonstrate the design features of the 606A concept in a shorter time with much lower costs. The resulting craft was named Avrocar and given the army designation VZ-9AV (ninth in a series of vertical take-off research aircraft). Most of the VZ-series aircraft looked like props from a James Bond movie.

The Avrocar was a saucer-shaped disk 18 ft. in diameter and 3 ft. thick. It was designed to go 300 mph and fly to an altitude of 10,000 ft. It weighed 5,650 lbs. and had separate cockpits for two crew members. Power was provided by a centrally located fan with a diameter of 5 feet. This was driven by the exhaust from three Continental J-69 turbojet engines. The flow from the fan was ducted to the periphery of the planform. An adjustable ring along the periphery was used to control the direction of the thrust. Two full-scale vehicles were built and were rolled out of the factory in May and August of 1959.

Ground tests of the full-scale propulsion system revealed there was not enough thrust available for hover out of the presence of the ground cushion. This was the first major problem with the program. The primary causes were large losses due to the complicated ducting of the flow and high internal temperatures that degraded the performance of the J-69s. All that could be done was to study the usefulness of the Avrocar as a ground-effect machine. At this point, the first Avrocar was sent to NASA Ames, Moffett Field, California, for wind-tunnel testing. Only here could its potential for forward flight (away from the ground) be assessed. The second vehicle remained at Avro for flight testing.

The first flight occurred on September 29, 1959. The Avrocar was tied to the ground for safety purposes. This flight lasted only 12 seconds, while the machine wobbled like a giant tiddlywinks. The first untethered flight occurred on November 12, 1959. These initial flights revealed a second major problem. At a height of 3 feet above the ground, uncontrollable pitching and rolling motions were encountered. The motion was termed *hubcapping*. The problem resulted from an erosion of the ground cushion as height was increased. Flight above this height was impossible.

Two formal Air Force flight evaluations were conducted at Avro, in

April 1960 and June 1961. During these tests, the vehicle reached a maximum speed of 35 mph. All attempts to control the hubcapping were unsuccessful. Meanwhile, the Ames wind-tunnel tests had shown that the VZ-9 had insufficient control for high-speed flight and was aerodynamically unstable. The addition of a conventional horizontal tail did not improve the situation. Thus, even if it could escape the ground cushion, the Avrocar would be unable to sustain high-speed flight.

Because the technical problems were insurmountable, the program was terminated in December 1961. A total of $10 million had been spent. One VZ-9 was scrapped; the other was given to the Smithsonian Institution. It now resides at the Smithsonian Air and Space Museum Annex in Silver Hill, Maryland.

The VZ-9 was the most ambitious flying-saucer research program, but not the only one. Many ground-cushion vehicle designs of the 1960s were saucer-shaped. The Convair Division of General Dynamics designed a 459-ft.-diameter vehicle with a gross weight of 4 million lbs. This monster was intended to span the oceans for naval operations. A nuclear reactor would provide the 150,000 hp necessary for operation. Obviously, it was never built. The remnants of these early efforts can be seen crossing the English channel today. Modern hovercraft use the same ground-cushion effect that gave early impetus to the flying-saucer research projects.

SOURCE MATERIAL

Genesis of the Program

Avro Aircraft Limited, 1959 (declassified 1971). *Pre-phase I Part I and Part 2 Development Program—VTOL Supersonic Aircraft.*

Technical Aspects

Avro Aircraft Limited, 1959 (declassified 1971). *The Avrocar Design.*

Greif, R. K., and W. H. Tolhurst, Jr. 1963. *Large Scale Wind-Tunnel Tests of a Circular Plan-Form Aircraft with a Peripheral Jet for Lift, Thrust, and Control.* NASA Technical Note D-1432.

Lindenbaum, B. L. 1990. *Historical Notes #9—Avrocar. Revolutions,* vol. 1, no. 6, published by the Archimedes Rotorcraft and VISTOL Museum, Brookville, Ohio.

Murray, D. C. 1990. *The AVRO VZ-9 Experimental Aircraft—Lessons Learned.* Presented at the AIAA Design, Systems, and Operations Meeting, September 1990, Dayton, Ohio.

Demise of the Program

Deckert, Lt. W. H., and Maj. W. J. Hodgson. 1962. *Avrocar Flight Evaluation.* Air Force Flight Test Center Report FTC-TDR-61-56, January.

Although it contains minor errors, the best generally available technical summary of the VZ-9 effort is given in:

Rogers, M. 1989. *VTOL Military Research Aircraft.* New York: Orion Books.

Part Two
THE CRASH AT ROSWELL

6. CRASH OF THE CRASHED-SAUCER CLAIM

Philip J. Klass

A revealing indication of the credulity of many of the present leaders of the UFO movement is their widespread acceptance of the claim that the U.S. government recovered one or more flying saucers in 1947, along with bodies of the alleged occupants—a tale rejected three decades ago by leading UFOlogists. A paper on the alleged crashed saucers was featured at the 1985 conference of the Mutual UFO Network (MUFON), the nation's largest UFO organization, and at earlier MUFON conferences.

The crashed-saucer tale was first advanced in 1950, barely three years after UFOs had been "discovered," in a best-selling book by Frank Scully, then a columnist for *Variety*—the "Bible of Show-Biz." But Scully's wild claim was promptly rejected even by *True* magazine, which itself had helped launch the UFO era a few months earlier when it published an article by Donald Keyhoe claiming that the earth was being visited by extraterrestrial craft.

Scully had obtained his information on the "crashed saucers" from two men who were exposed as con men two years later by a young reporter, J. P. Cahn, in an article published in *True*. Soon afterward, the two men were arrested and charged with selling a device called a "Doodlebug," which they claimed could find oil deposits. One of their victims had invested more than $230,000. The two men subsequently were convicted of fraud.

For almost three decades the claim of crashed saucers in New Mexico was ignored by responsible UFOlogists. Then, in 1980, it was resur-

---

This article originally appeared in the *Skeptical Inquirer* 10 (Spring 1986). Reprinted with permission.

rected by Charles Berlitz and William L. Moore in their book *The Roswell Incident.* Berlitz earlier achieved fame and fortune with his book on the Bermuda Triangle, which he claimed mysteriously swallowed up airplanes and ships—some of which had never existed. Moore earlier had authored the book *The Philadelphia Experiment,* which claimed that during World War II the U.S. Navy had discovered techniques that could make its ships invisible. But, according to Moore, the navy decided not to deploy this remarkable technique because its use gave sailors headaches or made them ill.

With this heritage, one might expect the leaders of the UFO movement to treat the Berlitz-Moore claims with considerable skepticism—unless one is familiar with the incredible credulity of many UFOlogists. Even Bruce S. Maccabee, one of the most technically competent of pro-UFOlogists and head of the Fund for UFO Research, gave the Berlitz-Moore book an endorsement in a book review published in *Frontiers of Science* magazine.

It is not surprising that Berlitz and Moore intentionally omitted from their book the considerable hard evidence that denied the claim of crashed saucers. But considering the amount of time UFOlogists spend in pouring over old, once-classified documents in a desperate search for evidence of a massive government coverup, it is curious that they too have failed to note, or publicize, how this utterly demolishes the crashed-saucer hypothesis.

According to Berlitz and Moore, a flying saucer crashed on the ranch of W. W. Brazel *during the first week of July 1947,* and possibly a second crashed near Socorro shortly afterward. The Army Air Force (soon to become the U.S. Air Force) position was that the debris found by Brazel was nothing more than a balloon-borne radar reflector, a device resembling a box-kite lined with aluminum foil, used to calibrate ground-tracking radars.

Naturally, Berlitz and Moore reject that explanation, drawing on the thirty-year-old recollections of local citizens and a number of newspaper clippings dating back to 1947. One important newspaper account Berlitz and Moore omit entirely is an Associated Press dispatch dated July 9, 1947, based on an interview with Brazel himself. The article quotes Brazel as saying he discovered the debris while riding his ranch on *June 14—more than two weeks before Berlitz and Moore claim the flying saucer crashed.*

Brazel's description of what he found, quoted in the Associated Press article, confirms the government position that the object was only a balloon-borne radar reflector: "large numbers of pieces of paper covered with a foil-like substance and pieced together with small sticks much like a kite. Scattered with the materials over an area of about two hundred yards were pieces of gray rubber. All the pieces were small." The article quoted Brazel as saying, "At first I thought it was a kite, but we couldn't put it together like any kite I ever saw."

According to Berlitz and Moore, the crashed saucer was promptly flown to Wright-Patterson Air Force Base, near Dayton, Ohio, for analysis. This base was the technical nerve-center for the air force and included its foreign intelligence operations. At the time, the base commander was Lt. Gen. Nathan Twining, who later became the USAF's chief of staff.

In September 1947, following a rash of UFO reports in the wake of the famous first sighting, reported by pilot Kenneth Arnold in June, the chief of staff of the Army Air Force had requested General Twining to provide him with a situation assessment, which Twining did in his letter of September 23, 1947. Berlitz and Moore quote extensively from this letter, including Twining's statement that "the phenomenon reported is something real and not visionary or fictitious." But the authors omit a critically important statement in the same letter, where Twining noted that there was a "*lack of physical evidence in the shape of crash-recovered exhibits which would undeniably prove the existence of these objects.*" And Twining was the commanding officer of the base where, according to Berlitz and Moore, top scientists had been analyzing the crashed saucer for more than two months.

After omitting this sentence from the Twining letter, the authors wrote: "it is understandable that the Twining memo makes no reference to the Roswell disc. . . ." It is understandable if the debris sent to Wright-Patterson AFB had turned out to be only a balloon-borne radar reflector and not a crashed saucer. The alternative explanations are that nobody thought to inform General Twining of the dramatic work under way at the base he commanded, or that Twining was intentionally lying to his own commanding officer.

Although dozens of ordinary citizens in New Mexico, without any official "need to know," quickly learned about the alleged crashed saucer(s), according to Berlitz and Moore word of the incident was withheld from the army chief of staff because, as they explain, "he did not possess the necessary clearances." His name: Dwight D. Eisenhower. Even after General Eisenhower became president, according to Berlitz and Moore, he was not informed of the recovered crashed saucer(s) until more than a year later because "some of the higher-ups in the intelligence community didn't trust Ike. . . ." (Recall that Allen Dulles, director of Central Intelligence under Eisenhower, was a brother of Secretary of State John Foster Dulles and a close personal friend of Eisenhower.)

In early 1953, top officials at Air Defense Command headquarters in Colorado Springs received a briefing on the USAF's UFO-investigations program by Capt. Edward J. Ruppelt, then head of Project Blue Book. The briefing was classified "Secret," as Ruppelt explained, in case sensitive matters, such as the coverage of the nation's air defense radar network, came up during the question-and-answer period. Subsequently, Ruppelt's prepared briefing was declassified and was published a decade ago in *Project Blue Book,* a book edited by Brad Steiger.

The head of Project Blue Book told top Air Defense Command officials: "It can be stated now that, as far as the current situation is concerned, there are no indications that the reported objects are a direct threat to the U.S., nor is there any proof that the reported objects are any foreign body over the U.S. or, as far as we know, the rest of the world. This always brings up the question of space travel . . . and it is the opinion of most scientists or people that should know that it is not impossible for some other planet to be inhabited and for this planet to send beings down to the earth.

"However there is no—and I want to emphasize and repeat the word *no*—evidence of this in any report the Air Force has received. . . . *We have never picked up any 'hardware.' By that we mean any pieces, parts, whole articles, or anything that would indicate an unknown material or object. . . .*"

Other hard evidence that denies the crashed-saucer claims can be found in material once classified "Secret" obtained from Central Intelligence Agency files in late 1978 via the Freedom of Information Act. These CIA papers reveal that in mid-1952, probably sparked by highly publicized reports of UFOs on radar screens at Washington's National Airport, the White House asked the CIA to make an independent assessment of the situation. As a result, high-ranking CIA scientists went to Dayton for a USAF briefing on the findings of its Project Blue Book effort. Then, in mid-August, these top CIA scientists briefed the director of Central Intelligence.

In one of these briefing papers, dated August 14 and originally classified "Secret," the briefer discussed the possible explanations for UFO reports, including the possibility that some might be generated by extraterrestrial craft. But the briefer added that "there is no shred of evidence to support this theory at present. . . . Another once-"Secret" briefing paper, dated August 15, states: "Finally, no debris or material evidence has ever been recovered following an unexplained sighting."

Recently, using the Freedom of Information Act, UFOlogists obtained an Air Intelligence Report, dated December 10, 1948, originally classified "Top Secret." It was considered such an important "find" that the *MUFON UFO Journal* devoted almost its entire July 1985 issue to reproducing this report, prepared jointly by the USAF's Directorate of

A HOAX UFO DOCUMENT

Don't be surprised if you see a tabloid headline that reads "TOP SECRET DOCUMENT REVEALS U.S. CONTACT WITH ETs," and a subhead that reads "Flying Saucer Under Test Since Early 1970s." A reduced-size copy of the document probably will be published, showing its bold "TOP SECRET" stamp at top and bottom.

The document is an obvious hoax, or should be an obvious hoax to any but a credulous UFOlogist and those who read sensationalist tabloids. The document is typewritten on a plain sheet of paper without the letterhead of any agency, such as the Department of Defense or the Central Intelligence Agency. It is undated and unsigned. Alongside each of the "TOP SECRET" stamp-marks is an "Unclassified" stamp-mark. But there is no notation to show when the document was "declassified" or the identity of the person who took the action—*which is mandatory even for "Secret" documents,* let alone those that are "Top Secret." (Such dates and annotations are found on the once-secret CIA documents dealing with UFOs that were released in late 1978.)

Even the contents are such conflicting nonsense that they reveal the document to be someone's practical joke. Following is the content, with XXX used here to indicate blacked-out portions:

> [Unclear letters/code?] PROJECT SIGMA: (PROWORD: XXX, Originally established as part of Project XXX in 1954. Became a separate project in 1976. Its mission was to establish communications with Aliens. This project met with positive success when in 1954, a USAF intelligence Officer met two Aliens at a prearranged location in the desert of New Mexico. The contact lasted for approximately three hours. XXXXXXXXXXXXXXXX the Air Force officer managed to exchange basic information with the two Aliens (Atch. 7). This project is continuing at an Air Force base in New Mexico (OPK: XXXXX
>
> [Unclear letters/code?] PROJECT SNOWBIRD: (PROWORD XXXX, Originally established in 1972. Its mission was to test fly a recovered Alien aircraft [*sic*]. This project is continuing [unclear word].
>
> [Unclear letters/code?] PROJECT XXXX XXXXXXX Originally established [unclear]. Its mission was to evaluate all UFO XXX information pertaining to space [unclear, several words].

According to the contents, Project Sigma, created as part of another project in 1954, was to "establish communications with Aliens," and in the same year, a USAF officer "met two Aliens . . . contact lasted for approximately three hours." Yet, despite this outstanding success, communications with Aliens was not considered important enough to warrant giving it separate project status until 22 years later, in 1976.

And, according to this document, the U.S. government either managed to borrow, or commandeered, a flying saucer by 1972, and seemingly has been flying it for more than a decade. Yet, curiously, none of this advanced "flying saucer technology" has found its way into latest-generation military aircraft, which still use "Old-fashioned" jet engines, wings, empennages, etc.

If the events described in this document had occurred, and if the U.S. government decided to declassify this document and release it to UFOlogists, knowing it would become public knowledge, why did not some American president hold a televised press conference to announce the startling news to the world?

It is interesting to note that the alleged contact with Aliens occurred in New Mexico, which of course is the state in which it is claimed that one or more flying saucers crashed in 1947. It's not really difficult to understand why there seems to be so much extraterrestrial traffic to that state, for, as many terrestrial visitors have discovered, New Mexico is true to its official motto: "Land of Enchantment."

Intelligence and the Office of Naval Intelligence. The objective of the report was to provide a best-estimate of the UFO situation as of 1948.

Although this once "Top Secret" report was prepared more than a year alter Berlitz and Moore claim that at least one crashed saucer was recovered in New Mexico by defense officials, *there is not a single mention of any such evidence.* Instead, the report focuses its speculation on the possibility that UFO reports might be generated by Soviet reconnaissance overflights, possibly using advanced vehicles built with the help of captured German scientists.

This 1948 report concludes: "IT MUST [*sic*] be accepted that some type of flying objects have been observed, although their identification and origin are not discernible. In the interest of national defense it would be unwise to overlook the possibility that some of these objects may be of foreign origin."

Presumably this report will be studied by MUFON's international director, Walter Andrus, and by many other leading UFOlogists. Will they recognize its obvious implications (and the other hard evidence cited above) in terms of the Berlitz-Moore claim of crashed saucers? Or will Moore continue to spin his tales at future MUFON conferences, prompting his audience to believe that somewhere in some secret government vault lies the debris, and perhaps even the bodies, that could at long last confirm UFOlogists' fondest hopes?

7. THE MJ-12 CRASHED-SAUCER DOCUMENTS

Philip J. Klass

On May 29, 1987, William L. Moore and two associates, Stanton Friedman and Jamie Shandera, released what purport to be "Top Secret" government documents that are either the biggest news story of the past two millennia or one of the biggest cons ever attempted against the public and the news media.

If authentic, the documents show that the U.S. government recovered a crashed flying saucer in mid-1947 and four extraterrestrial-creature bodies, such as Moore claimed in his 1980 book, *The Roswell Incident* (coauthored with Charles Berlitz), and that the government also recovered the remains of another saucer, which crashed on December 6, 1950, near the Texas-Mexico border.

Further, these documents indicate that on September 24, 1947, President Harry S. Truman authorized Defense Secretary James Forrestal and Dr. Vannevar Bush, president of the Carnegie Institution, to create a top-secret panel of twelve scientists, military leaders, and intelligence officials—called Operation Majestic-12 (MJ-12). Its function, presumably, was to analyze the crashed saucer to determine its technological secrets and to make recommendations for a suitable U.S. response to extraterrestrial visitors whose intentions might prove to be hostile.

The papers released by Moore, Friedman, and Shandera consisted of three elements, purporting to be the following: (1) a "Top Secret" memorandum from President Truman to Defense Secretary Forrestal, dated

---

This article originally appeared in the *Skeptical Inquirer* 12 (Winter 1987–88). Reprinted with permission.

September 24, 1947, authorizing him and Dr. Bush to proceed with Operation Majestic-12; (2) a seven-page "Top Secret/Eyes Only" Majestic-12 document used to brief President-Elect Eisenhower, dated November 18, 1952; (3) a "Top Secret" memorandum from Robert Cutler, special assistant to President Eisenhower, to General Nathan Twining, USAF chief of staff, dated July 14, 1954.

According to Moore, the Truman/Forrestal memo and the Eisenhower briefing document were received in mid-December 1984 by Moore's friend Jamie Shandera, a Los Angeles television writer-producer, on an undeveloped roll of 35 mm film.

As Moore described the circumstances in his banquet speech at the 1987 MUFON UFO conference in Washington in late June, the package containing the film was wrapped in plain brown paper "taped with official-looking brown tape on all seams. The address label was carefully typed, with no return address. Inside the [brown] wrapper was a second one, similarly sealed, inside of which was yet another white envelope, inside of which was a canister, inside of which was a roll of *unprocessed* film." (Moore has not replied to my repeated requests that he send me a photocopy of the postmark, showing city and date of mailing.)

If the MJ-12 documents film is authentic, it is odd that it was not sent to Moore, whose book and numerous MUFON conference papers have made him world famous as *the* leading crashed-saucer proponent and researcher—or to Stanton Friedman, who has been Moore's closest collaborator on crashed-saucer research for almost a decade. As Moore explained at the MUFON conference, in recent years he has focused his efforts on trying to establish contacts within the intelligence community "to find out what happened to the wreckage after it came into custody of military authorities."

Why would the film be sent to Shandera, who had never published any papers on UFOs or crashed saucers and does not even consider himself a UFOlogist? How would the sender of the 35 mm film even know that Shandera and Moore were friends and that the contents would find their way to Moore?

Even before the film was developed and the MJ-12 papers became visible, Shandera demonstrated "psychic powers" in "knowing" that the undeveloped roll of 35 mm film in the plain brown wrapper from an unknown sender would be of interest to Moore. This explains why he promptly called Moore even before the film was processed and why Moore was present when it was being developed, according to Moore's report to MUFON.

According to Moore, the person who made the 35 mm film had photographed the MJ-12 documents in two duplicate sequences, seemingly to ensure that there would be at least one good set of imagery. But the sender had not thought to process the film himself for final assurance before sending it to Shandera.

The film's seven-page Eisenhower briefing document indicated that the briefing officer was Admiral Roscoe H. Hillenkoetter, who had been head of the Central Intelligence Agency in 1947 when MJ-12 allegedly was created and thus would logically be a member. But in the fall of 1950 Hillenkoetter left the CIA to return to sea duty as commander of the Seventh Task Force in Formosan Waters and did not return for duty in the United States until late 1951—the year before the alleged briefing—to become commander of the Third Naval District in New York.

It would have been more logical for Eisenhower to have been briefed by the chairman of MJ-12, who had remained in the United States, close to the committee's activities, since 1947. Presumably this would have been Dr. Bush, who allegedly organized MJ-12 and is shown as one of its original members. (Although the briefing document lists the 12 members of the group, it does not indicate who was chairman but identifies Hillenkoetter as "MJ-1.")

While there are many such substantive anomalies in the contents of the alleged Hillenkoetter/ Eisenhower briefing documents, which will be discussed in a subsequent article,* the most revealing is the format used to write dates. (I am indebted to Christopher Allan, Stoke-on-Trent, England, who first brought these very significant anomalies to my attention.)

Whoever typed the Hillenkoetter briefing document used a peculiar style for writing dates—an erroneous mixture of civil and military formats. In the traditional civil style, one would write: November 18, 1952. Using the standard military format, one would write: 18 November 1952. But whoever typed the Hillenkoetter briefing document used a military format with an unnecessary comma: "18 November, 1952." *Every date* that appears in this document uses this erroneous military format, with the "unnecessary comma." By a curious coincidence, this is precisely the same style used by William L. Moore in *all* of his many letters to me since 1982, when our correspondence began.

Another curious anomaly in the Hillenkoetter document is the use of a "zero" preceding a single-digit date, a practice that was not used in 1952, when the briefing document allegedly was written, and which has come into limited use only in very recent years. Examination of numerous military and CIA documents written during the 1950s, 1960s, and 1970s shows the standard format was to write: "1 August 1950." Yet the Hillenkoetter document contains the following: "01 August, 1950" and "07 July, 1947," and "06 December, 1950."

---

* "MJ-12 Papers Authenticated," *Skeptical Inquirer* 13 (Spring 1989): 305–309.

My files of correspondence from Moore show that he used a single digit without a zero until the fall of 1983—roughly a year before the Hillenkoetter document film reportedly was sent to Shandera—when he then switched to the same style used in the Hillenkoetter briefing document.

The other document contained on the 35 mm film is what purports to be a "Top Secret/Eyes Only" memorandum, dated September 24, 1947, on White House stationery signed by President Truman. There is no question of the authenticity of the signature, but thanks to invention of the Xerox machine, it is easy to substitute bogus text on a photocopy of an authentic original, obtained, for example, from the Truman Library, in Independence, Missouri, which both Moore and Friedman visited prior to late 1984.

The format of the September 24 memorandum to Defense Secretary Forrestal differs significantly from that used by the president's secretary in other memoranda written to Forrestal, and others, during the same period. The typewriter used for the September 24 document was a relatively inexpensive one with a worn ribbon and keys that had not been recently cleaned, in contrast to the more elegant typeface, fresh-ribbon appearance of authentic Truman memoranda written at about the same time.

Furthermore, Truman was a blunt-spoken man whose letters reflect that style. Yet the second paragraph of the two-paragraph September 24 memo is filled with "un-Truman-like" gobbledygook: "It continues to be my feeling that any future considerations relative to the ultimate disposition of this matter should rest solely with the Office of the President following appropriate discussions with yourself, Dr. Bush, and the Director of Central Intelligence." There was no need for Truman to be vague for security reasons, because the September 24 letter is stamped "Top Secret/Eyes Only."

If the letter were authentic, I'm confident it would have read more like the following: "Let's find out where in the hell these craft are coming from, whether they pose a military threat, and what in the hell we can do to defend the country against them if they should attack. I trust you will place all our forces on alert status and inform me if you need additional funds or other resources to protect this nation."

Moore told his MUFON audience that for two and a half years "we sat on the [MJ-12] material and did everything we could with it" to check its authenticity. He noted that all of the persons listed as being members of MJ-12 are now dead. Moore added: "If I was going to pick a panel at that time, capable of dealing with a crashed UFO, I would certainly want to consider [those on] that list." In other words, the members of MJ-12 were persons whom Moore himself would probably have selected for such a committee.

In mid-1982, more than two years before learning of Bush's key role from the MJ-12 papers, Moore demonstrated remarkable psychic abilities in a paper presented at a MUFON conference in Toronto. Moore said that Bush would be "the logical choice for an assignment to set up a Top Secret project dealing with a crashed UFO." Two years later, the MJ-12 papers confirmed Moore's judgment.

In the spring of 1985, Friedman learned that more than a hundred boxes of once Top Secret USAF intelligence documents from 1946 through 1955 were being reviewed by USAF representatives for declassification at the National Archives, in Washington, and he informed Moore of this. In July, Moore and Shandera flew to Washington and were the first persons—according to Moore—to gain access to those more than one hundred cartons of once Top Secret documents.

Lady Luck smiled, enabling Moore and Shandera to discover a sorely needed sheet of paper that could authenticate the MJ-12 documents on the 35 mm film. This key document purports to be a brief, two-paragraph memorandum, dated July 14, 1954, to USAF Chief-of-Staff Twining written by Robert Cutler, then special assistant to President Eisenhower. The subject of Cutler's memo was "NSC/ MJ-1 2 Special Studies Project," and it informed General Twining that "the President has decided that the MJ-12 SSP briefing should take place during the already scheduled White House meeting of July 16, rather than following it as previously intended."

Moore explained the importance of the July 1985 discovery of the Cutler memo to his MUFON audience in these words: "For the first time we had an official document available through a public source [National Archives] that talked about MJ-12." One might logically have expected that Moore would promptly "go public" with his remarkable MJ-12 papers, which now seemingly were authenticated by the Cutler memo. Yet, curiously, Moore did not do so, *for nearly two years!*

In the April 30, 1987, issue of a newsletter Moore publishes, he first released three of the seven pages of the Hillenkoetter briefing document, but in heavily censored form—*censored by Moore himself.* There was no mention of the Truman memo of September 24, 1947, nor of the Cutler memo of July 14, 1954, nor of the 35 mm film. Instead, Moore implied that the three heavily censored pages of the Hillenkoetter document had been provided by his "well-placed contacts within the American intelligence community" and said that "assurances have been given that additional information can be made available to us over the next several months."

This suggests to me that Moore planned to "dribble out" the MJ-12 material, in his possession since late 1984, in subsequent issues of his newsletter. This could generate more paid subscribers. If this was Moore's plan, it was thrown into disarray in mid-May when British

UFOlogist Timothy Good met with the press to promote his new book, which claims a global UFO coverup. Good told British news media about the MJ-12 documents, which he said he had obtained "two months ago from a reliable American source who has close connections with the intelligence community. . . ."

Shortly afterward, Moore went public with the MJ-12 documents, including the Truman and Cutler memoranda, crediting them to the Moore-Shandera-Friedman Research Project. His release said: "Although we are not in a position to endorse its authenticity at this time, it is our considered opinion, based upon research and interviews conducted thus far, that the document and its contents *appear* to be genuine. . . . One document was uncovered at the National Archives which unquestionably verifies the existence of an 'MJ-12' group in 1954 and definitely links both the National Security Council and the president of the United States [Eisenhower] to it. A copy of this document, with its authenticating stamp from the National Archives, is also attached for your examination."

Stanton Friedman, nuclear physicist turned full-time UFO lecturer, who recently has returned to his original field, has been Moore's principal researcher-collaborator on crashed-saucer matters. Moore and Friedman continued to collaborate even after Friedman moved from California to New Brunswick, Canada, in 1980, as evidenced by their jointly authored paper on crashed-saucers presented at the 1981 MUFON conference in Cambridge, Massachusetts.

Thus one would think that, immediately after discovering the MJ-12 papers on the 35 mm film in late 1984, Moore would have sent a copy to Friedman. Yet it was not until late May 1987 that Friedman obtained a set of the documents, according to Friedman. In view of Moore's claim that he and Shandera spent more than two years trying to verify the authenticity of the MJ-12 papers, one would have expected that Moore would promptly have sent the MJ-12 papers to Friedman to enlist his help in trying to authenticate them.

Friedman told me that Moore first informed him by telephone of the MJ-12 papers in late 1984 or early 1985. But, as Friedman explained in a recent letter, at that time one of his sons was fatally ill and Friedman was preoccupied with buying a new house and preparing to leave for a long UFO lecture tour. So it did not occur to Friedman to ask that Moore send him a copy of the MJ-12 papers, nor did he request a copy during the subsequent two years. That Moore did not send Friedman a copy on his own seems a most curious oversight since the documents, if authentic, were world-shaking in their importance. It is especially odd in view of Timothy Good's claim that his unidentified American source had supplied him with a copy of the MJ-12 papers earlier than Moore supplied a copy to

Friedman, his closest collaborator on the case. (Moore has not responded to my repeated queries as to whether he was the American source who supplied Good with the MJ-12 papers.)

On July 22, 1987, Jo Ann Williamson, chief of the military reference branch of the National Archives, wrote a three-page memorandum summarizing the results of its own investigation into the Cutler/Twining memo, which played a key role in "authenticating" the MJ-12 papers. The National Archives memo pointed out that every other Top Secret document in the boxes of material in which the Cutler/Twining memo allegedly was found was stamped with an individual "register number"—a protocol used by the USAF reviewers to assure that each is properly accounted for and none is mislaid. The National Archives memo notes that the Cutler/Twining memo "does not bear such a number."

The Cutler/Twining memo purported to be a carbon copy on onionskin paper—which understandably would not carry the White House logo and would not necessarily be signed or initialed by Cutler. But the National Archives memo noted that "the Eisenhower Library has examined its collection of the Cutler papers. All documents created by Mr. Cutler while he served on the NSC staff have an eagle watermark in the onionskin paper." The Cutler/Twining memo found by Moore and Shandera did *not* have such a watermark. Furthermore, typewriter-key impressions protruded from the backside, suggesting it was an "original" and not a carbon copy as it appeared to be.

The National Archives memo quoted Eisenhower Library officials as stating that even when President Eisenhower had "off-the-record" meetings, his appointment books "contain entries indicating the time of the meeting and the participants. . . ." But "President Eisenhower's Appointment Books contain no entry for a special meeting on July 16, 1954, which might have included a briefing on MJ-12."

More significant, Robert Cutler could not possibly have written the memo on July 14, 1954, telling of last-minute changes in the president's schedule, *because Cutler had left Washington 11 days earlier (July 3)* to visit major military facilities in North Africa and Europe and did not return to Washington until July 15. This is shown by his subsequent trip report to the president, dated July 20, housed in the Eisenhower Library.

On August 20, 1987, the Committee for the Scientific Investigation of Claims of the Paranormal issued a four-page press release that characterized the MJ-12 papers as "clumsy counterfeits." It cited some of the discrepancies discussed above and attached a copy of the National Archives memo of July 22, 1987.

Several weeks later, the J. Allen Hynek Center for UFO Studies responded with a press release that said the question of the authenticity of the MJ-12 documents "is still open." CUFOS quoted Moore as saying that CSICOP "failed to raise a single issue which cannot be explained by further examination of the evidence." Moore charged that CSICOP's appraisal was "not only premature, but unscientific and emotional."

Shortly afterward, Citizens Against UFO Secrecy, a group that often accuses the government of a UFO coverup, distributed the September issue of its newsletter *Just Cause*. The entire issue was devoted to the MJ-12 papers. Editor Barry Greenwood said that he remains open-minded to the possibility that a flying saucer crashed in New Mexico in 1947. But, based on his own investigation into the MJ-12 papers, Greenwood characterized them as "a grand deception and, consequently, a giant black eye on the face of UFOlogy. . . . The deeper we looked, the worse it became."

8. MJ-12 PAPERS "AUTHENTICATED"?

Philip J. Klass

"Linguistics Expert Vouches for MJ-12 Briefing Paper" was the headline in the *MUFON UFO Journal. The International UFO Reporter,* published by the Hynek Center for UFO Studies (CUFOS), headlined its article "MJ-12 Document Authentic, Says Expert." *UFO* magazine's headline was "Linguistic Analysis: MJ-12 Document Validated."

This disputes my own findings that the "Top Secret/Eyes Only" documents—which seemingly reveal that the U.S. government recovered two crashed flying saucers and the bodies of four UFOnauts in 1947 and 1950—are counterfeit, for the many reasons detailed in *Skeptical Inquirer.* (See Winter 1987–88: 137–46 [chapter 7 in this volume]; Spring 1988: 279–89.)

The newsletter *Focus,* published by William L. Moore, who released the MJ-12 documents, which seemingly confirm claims made in a book he coauthored in 1980, headlined its article: "MJ-12 Document Is Real, Says Expert."

The "expert" is Roger W. Wescott, professor of linguistics at Drew University in Madison, New Jersey, whose vita suggests he should be well qualified for such an assignment. Wescott also has a longstanding interest in a broad spectrum of the paranormal, including UFOs, which could explain why he was selected to make a linguistic analysis of the MJ-12 papers by Robert H. Bletchman, MUFON's state director for Connecticut.

Wescott finds the popular extraterrestrial-craft explanation for UFOs too prosaic for his taste. Instead, as he later explained to me, he sees a

---

This article originally appeared in the *Skeptical Inquirer* 13 (Spring 1989). Reprinted with permission.

direct connection between UFOs and "these things that have been around for centuries [such as] fairy phenomena, wee folk, strange events of all kinds, strange appearances that baffle people."

Wescott spent a total of eight hours on his analysis, for which he was paid $1,000—jointly provided by MUFON (Mutual UFO Network), CUFOS, Fund for UFO Research, and Moore's own "Fair Witness" organization.

The principal portion of the MJ-12 papers is what purports to be a Top Secret/Eyes Only document used by Rear Adm. R. H. Hillenkoetter to brief President-elect Dwight D. Eisenhower on November 18, 1952, on the history of the so-called Top Secret Majestic-12 Committee. This committee allegedly had been created by President Harry S. Truman on September 24, 1947, to analyze the crashed saucers and alien bodies and to cope with resulting national defense issues.

Hillenkoetter had been director of the Central Intelligence Agency in mid-1947, when the first crashed saucer allegedly was recovered. He held that post until the fall of 1950, when he returned to the Navy and was assigned a post in the Pacific. If the MJ-12 papers are to be believed, Hillenkoetter not only continued as a member of MJ-12 during his Pacific duty but was selected to brief President-elect Eisenhower.

On November 1, 1987, after I learned that Wescott had been approached by Bletchman, I sent him several white papers, pointing out what seemed to me to be serious discrepancies that indicated the documents were counterfeit.

The most important of these focused on a stylistic issue that I expected would especially interest Wescott. The alleged Hillenkoetter briefing document *consistently* used an extremely unusual mixed military-civil format for writing a date. The format typically used by civilians, for example, is "November 18, 1952" while the military format would be "18 November 1952."

But the MJ-12 briefing paper *consistently* used a mixed format with a superfluous comma, for example, "18 November, 1952." Additionally, when there was a single-digit date, the MJ-12 document had a zero before the digit, i.e., "07 July, 1947." This style was not used in the United States in the early 1950s, when the document allegedly was written.

I also sent Wescott a white paper that revealed that William L. Moore *consistently* used this same unusual format, with a "superfluous comma" and a "preposed zero" before a single-digit date. My paper provided photocopies of thirteen examples from Moore's personal letters to me with superfluous comma and preposed zero underlined.

A critical question was whether Hillenkoetter also used this mixed military-civil date format prior to November 18, 1952, when the briefing document was allegedly prepared. At my request, the Truman Library provided me with four letters Hillenkoetter had written to President Truman in 1948–1950 during his tenure as CIA director.

Every one of these genuine Hillenkoetter letters/memoranda used the traditional military date format, *without a superfluous comma.* Three of the four were written on single-digit dates but *none used the preposed zero* found in the MJ-12 document.

To the best of my knowledge, the only two examples of the consistent use of this mixed military-civil format for writing the date and a preposed zero are William L. Moore's letters and the alleged Hillenkoetter briefing document.

In early 1988, Stanton T. Friedman, Moore's longtime collaborator, who has strongly endorsed the MJ-12 papers, visited the Truman Library to obtain copies of Hillenkoetter letters/memoranda so that he could give them to Wescott for his comparison of their style-format with that of the MJ-12 papers.

Friedman later provided me with copies of sixteen additional Hillenkoetter letters/memoranda written between 1947 and 1950, before he returned to sea duty. *Every one of these uses the conventional military date format, i.e., without a superfluous comma.* Four of these were written on single-digit dates but *none of these used the preposed zero found in the MJ-12 documents.* Additionally, every one of these authentic Hillenkoetter letters/ memoranda showed the writer's name as "R. H. Hillenkoetter," whereas the MJ-12 papers refer to the briefer as "Roscoe H. Hillenkoetter."

For Wescott's linguistic analysis of the MJ-12 papers, he was supplied with a total of twenty-seven Hillenkoetter documents, including those he wrote as CIA director as well as private letters written after he had retired.

Wescott told Bletchman he would make his assessment based on "stylistics"—a discipline of linguistics that deals with the more or less unique design and syntax characteristics of a person's written language. On April 3, 1988, Wescott wrote Bletchman to render his verdict. Wescott's letter revealed that he had misunderstood the issue of the mixed military-civil date format and superfluous comma that I had earlier raised and documented for him.

Wescott said: "The stylistic evidence that [Klass] cites seems to me to be quite inconclusive: I myself, for example, alternate between writing 'April 3, 1988' and '3 April 1988' in my own letters." He added: "In ambiguous situations like this, I tend to follow an equivalent of the legal principle 'innocent till proven guilty.' My analog is 'authentic till proven fraudulent.' "

Four days later, on April 7, 1988, Wescott again wrote to Bletchman to say that Stanton Friedman had just called, seeking a less ambiguous endorsement of MJ-12 authenticity. This motivated Wescott to offer the following endorsement: "In my opinion, there is no compelling reason to regard *any* of these communications as fraudulent or to believe that any of them were written by anyone other than Hillenkoetter himself. This statement holds for the controversial presidential briefing memorandum of November 18, 1952, as well as for the letters, both official and personal."

I couldn't believe my eyes when I read the foregoing in the *MUFON UFO Journal.* The 27 unquestioned, authentic Hillenkoetter letters/memoranda had been supplied to Wescott to provide a stylistic benchmark for appraising the authenticity of the MJ-12 papers. But judging from Wescott's statement, seemingly he spent some of his eight hours in assessing their authenticity. It is not clear what he used as a benchmark for this process.

Wescott sent me a copy of his letter of May 15, 1988, to Mark Rodeghier, scientific director of CUFOS, thanking him for payment and offering additional views on MJ-12. In this letter, Wescott mentioned the "mixed military-civil format" but again completely failed to grasp the obvious stylistic issue involved.

Commenting on the preposed zero before single-digit dates, which I claimed had not come into use until the 1970s, Wescott said: "If it is like most other matters of style and usage, I would say, it came in gradually and sporadically rather than suddenly and systematically." The critical issue was *when* did the preposed zero first begin to come into use in the United States.

On May 23, I wrote Wescott and asked him to supply me with photocopies of five U.S. military or CIA documents written prior to the MJ-12 document date that used the preposed zero in one-digit dates. To provide additional incentive, I offered to contribute $100 to his favorite charity for each such letter he provided, up to a maximum of $500. On June 18, having failed to hear from Wescott, I wrote him and raised the ante. I offered to contribute $100 per letter for up to ten letters, or a total of $1,000.

After a month passed without a response from Wescott, I wrote to make an additional offer: For each authentic Hillenkoetter letter/ memoranda dated prior to November 18, 1952, that used a preposed zero and bore the name "Roscoe H. Hillenkoetter" (rather than "R. H. Hillenkoetter") I would contribute $200 to Wescott's favorite charity, up to a maximum of $2,000.

Thus, if Wescott had any hard evidence to support his claim, he could obtain as much as a $3,000 contribution from me for his favorite charity simply by sending me photocopies of any such documents. Wescott never replied to any of these offers.

By early October, I had written Wescott six letters to which he had never replied, the last being on August 30, so I decided to call him. I reached him in Chattanooga, where he now lives, having accepted a two-year assignment at the University of Tennessee as the "first holder of the endowed chair of excellence in the humanities."

In early correspondence, Wescott had written that in his examination of the MJ-12 papers he had found no "clear evidence of fraud," prompting me to ask for illustrative examples of what he would consider to be "clear evidence of fraud." Wescott replied: "If someone were to come forward and confess fraud and then could show the means by which the fraud was perpetrated, that would be relatively conclusive."

When I asked Wescott, who is sixty-three, how many documents of questionable authenticity he had analyzed during his long career, he replied: "A small number . . . several." He added that authentication "isn't something that I usually do." Wescott said, "The Hillenkoetter documents are the first in which I was asked to do anything official." He explained that in the other instances he had not conducted an analysis and had simply been asked for his "impressions" as to the document's authenticity. Wescott added, "This is not my specialty."

On June 10, 1988, Wescott had sent out a form letter addressed to "Dear Colleagues" to thank those who had written about his then recent assessment of the MJ-12 papers. He admitted that he had "stepped into a hornet's nest of controversy."

"On behalf of those who support the authenticity of the memo, I wrote that I thought its fraudulence unproved," Wescott wrote. "On behalf of its critics, *I could equally well have maintained that its authenticity is unproved*" (emphasis added). *But he opted not to do so.* The question of crashed saucers, Wescott wrote, "like the larger 'ufological' topic of which it is a part, *will remain to perplex us, I suspect, for a long time*" (emphasis added).

During my telephone conversation with Wescott in October, I asked if he agreed that "if the MJ-12 papers are authentic, it indicates the most extraordinary event of at least the last two millennia?" Wescott replied: "Oh no, I don't think I would go that far." I was surprised at his reply and noted that if the documents were authentic then the United States would have solid proof of extraterrestrial visitations. Wescott replied: "They wouldn't have to be *extraterrestrial.* They could be what's called '*ultra-terrestrial.*' " When I sought a clarification of the latter term, Wescott explained: "Meaning they didn't come from outside the earth. . . . Another possibility is that simply there are more dimensions to our existence than we understand and that occasionally there are interferences from one domain to another."

In one of Wescott's very few responses to my letters, he wrote on May 13 to say that he was "not as impressed by CSICOP and the *Skeptical Inquirer* as you, because I don't find them genuinely skeptical." Instead he characterized them as "counterfaith."

The foregoing should provide a useful perspective for readers who chance to read an article that cites Wescott's endorsement of MJ-12 authenticity, such as *UFO* magazine's article. It began: "After eight hours of stylistic analysis, noted linguistics expert Dr. Roger W. Wescott has offered what can be considered the first professional authentication of . . . MJ-12 documents. . . ." The magazine quoted Moore as commenting that Wescott is "saying flat out that in his opinion . . . Hillenkoetter wrote it."

The *International UFO Reporter* (*IUR*) article began: "After comparison with letters and other materials known to have been written by Adm. Roscoe Hillenkoetter, Roger W. Wescott . . . has concluded that the much-disputed MJ-12 document was composed, as claimed, by Hillenkoetter. A later issue of *IUR* carried Wescott's more equivocal assessment of June 10, under the headline: "Statement from Roger Wescott." There was no CUFOS comment or reference to the earlier *IUR* claim that Wescott had authenticated MJ-12.

Considering that the MJ-12 papers represent Wescott's first "official" role in trying to assess the authenticity of a document of great potential importance, some might expect he would write a paper for an appropriate journal. But when he was asked about this possibility, he said he had no such intentions.

Under the circumstances, that is not surprising.

9.
NEW EVIDENCE OF MJ-12 HOAX
Philip J. Klass

A "smoking gun" recently has been discovered that confirms beyond any doubt that the alleged "Top Secret/Eyes Only" MJ-12 documents, which seemingly showed that the U.S. government had captured at least one crashed flying saucer and the bodies of several extraterrestrials in 1947, are counterfeit.

The MJ-12 documents were made public on May 29, 1987, by William L. Moore and two associates, Jaime Shandera and Stanton T. Friedman. If authentic, the documents would confirm claims made in a 1980 book, *The Roswell Incident,* authored by Moore and Charles Berlitz, of "Bermuda Triangle" fame.

The MJ-12 papers include what purports to be a one-page memorandum from President Harry Truman to Defense Secretary James Forrestal, dated September 24, 1947—several months after the alleged crashed-saucer recovery in New Mexico. The letter authorized Forrestal and Vannevar Bush to create a top-level Majestic-Twelve (MJ-12) group to analyze the crashed saucer and alien bodies. The other MJ-12 document is a lengthy status report on MJ-12's crashed-saucer research efforts, seemingly intended to brief President-elect Eisenhower, dated November 18, 1952. The briefing paper seemingly was written by Rear Admiral R. H. Hillenkoetter, who had earlier headed the Central Intelligence Agency and allegedly was a member of MJ-12.

A roll of 35 mm film, together with photocopies of these two "Top Secret/Eyes Only" documents, reportedly arrived at the home of Shan-

---

This article originally appeared in the *Skeptical Inquirer* 14 (Winter 1990). Reprinted with permission.

dera by mail from an unknown sender on December 11, 1984. Moore, Shandera, and Friedman claim that they spent the next two and a half years investigating the authenticity of the MJ-12 papers before making them public in May 1987.

Moore and his associates said that their lengthy investigation had failed to turn up anything that would cast doubt on the authenticity of the MJ-12 papers. My own investigation revealed many reasons to suspect the MJ-12 papers were counterfeit. (See my two articles published in *SI*: Winter 1987–88, p. 137; Spring 1988, p. 279.)

Recently, I discovered hard physical evidence that demonstrates that these documents are counterfeit. This is based on the fact that a person's handwritten signature is like a snowflake—no two are ever *identical.*

Before the advent of the "Xerox Era" and "signature-machines," the very existence of two identical signatures was considered to be "*very strong evidence of forgery,*" according to the book *Questioned Documents,* by Albert S. Osborn, published in 1978. Osborn notes that "the fact that two signatures are very nearly alike is not alone necessarily an indication of forgery of one or both but the question is whether they are *suspiciously alike*" (emphasis added).

The "Harry Truman" signature on the MJ-12 Truman memorandum of September 24, 1947, is *suspiciously like* the signature on the letter that Truman wrote to Vannevar Bush on October 1, 1947, the original of which I found in the Bush collection in the Manuscript Division of the Library of Congress and made several photocopies of it there.

In signing the authentic letter to Bush, Truman's pen accidentally skidded slightly, creating a small extraneous mark on the left upper part of the right-hand vertical stroke in the letter "H." *The same "skidmark" appears on the Truman signature of the MJ-12 memo of September 24, 1947.* It is slightly heavier on the MJ-12 memo because of the multiple photocopying operations used to make the hoax document.

(Photocopies of both signatures are shown on the opposite page. Readers who are sufficiently interested can make photocopies and superimpose them before a strong light to confirm that the two are identical.)

If the Truman signature is a counterfeit, then so is the alleged Hillenkoetter MJ-12 briefing paper, contained on the same 35 mm film, which makes specific reference to this "special classified executive order of President Truman on 24 September, 1947. . . ."

To obtain an expert corroboration of my own findings, I called David Crown, a professional "document examiner" in the Washington, D.C., area, who previously headed the Central Intelligence Agency's questioned documents laboratory. Crown informed me that the Truman memo had already been exposed as a hoax because it was written on a typewriter

Dr. Bush:

I appreciated very much your good
of September twenty-sixth and I hope
; will work out in a satisfactory manner
oming season.

Sincerely yours,

o be my feeling that any future
tive to the ultimate disposition
ld rest solely with the Office
llowing appropriate discussions
Bush and the Director of Central

4.88

7.95

2.97

2 50

2.40

1.72

4.41

6.13

3.2% longer than signature on Truman-Bush.

that "did not even exist in 1947." He told me that this discovery had been made by a highly respected document examiner, whose name and telephone number he provided. (I will refer to the latter document examiner as PT because of his reluctance to become a public figure in the MJ-12 controversy.)

When I called PT, he expressed great interest in obtaining a copy of the authentic Truman-Bush signature of October 1 because he had earlier been drawn into the MJ-12 controversy through a friend, also a professional document examiner. PT's earlier analysis of the typeface of the machine used to prepare the MJ-12 Truman memo indicated that it was a Smith-Corona machine that first appeared in 1963—more than fifteen years *after* the September 24, 1947, date on the memo.

PT asked me to send the October 1 memo to him by overnight mail because he was leaving in two days for a meeting of professional document examiners in San Francisco, and I did so. In our first conversation, I mentioned that the MJ-12 Truman signature was approximately 3.6 percent longer than the one on the October I letter, which I attributed to optical distortion during the several photocopying operations needed to produce a counterfeit. PT explained that Xerox, and its competitors, intentionally do not reproduce a thin border around the outside of a document to be copied—to avoid creating unwanted lines at the edges. To compensate for this, the original copy is enlarged by roughly 1.2 percent—which is imperceptible to the casual reader.

Thus, if a counterfeiter had needed three photocopying iterations to produce the MJ-12 memo—as my own experiments suggested—this would account for the fact that the MJ-12 signature is about 3.6 percent larger than the October 1 signature.

Eight days later, PT called and informed me that the MJ-12 signature was "a classic signature transplant," i.e., a photocopy forgery. In the authentic October 1 signature, a portion of the top of the "T" in "Truman" barely intersected the "s" at the end of "Sincerely yours." When the counterfeiter had used typewriter correction fluid to retouch out the "Sincerely yours," he had slightly "thinned" the width of the top of the "T." This retouching, PT told me, is the "kind of coup de grace we look for." PT told me he had made overhead projector transparencies of the MJ-12 and October I signatures and taken them to San Francisco to show at the meeting of professional document examiners. He first showed his audience the MJ-12 Truman memo typeface, pointing out that the Smith-Corona machine used did not exist in 1947. Then PT showed the MJ-12 Truman signature and superimposed a copy of the October 1 signature—enlarged by about 3.6 percent—and pointed out the "thinning" of the top of the "T." PT said his audience gave a verbal endorsement—"a chorus of 'Ah-haa!' "

PT told me he had already called Moore's longtime associate Stanton Friedman to inform him of PT's findings because "he had [earlier] sent me all this [MJ-12] material . . . [and] I felt I owed it to him to tell him that he should just wash his hands of this." (Friedman opted to ignore PT's advice. The next week Friedman spoke at a MUFON regional conference near St. Louis and repeated his earlier endorsement of the authenticity of the MJ-12 papers.)

Friedman, who has been the most outspoken defender of the authenticity of the MJ-12 papers, knew at least shortly after their release—more than two years ago—that the Truman signature on the MJ-12 memorandum "match[ed]" the one on a letter Truman wrote to Bush in October 1947.

Friedman reported this fact in his article published in the September/October 1987 *International UFO Reporter* claiming that this "match" confirmed the authenticity of the MJ-12 document. In fact, it really revealed just the opposite. (I am indebted to Christopher D. Allan of the United Kingdom for bringing Friedman's claim to my attention, and to Joe Nickell for supplying references from the book *Questioned Documents.*)

Earlier this year, Friedman requested and received a $16,000 grant from the Fund for UFO Research (FUFOR) for further investigation into the authenticity of the MJ-12 papers. Ironically, he already had in his possession the "smoking gun." Friedman, in an interim report on his FUFOR funded research, published in the September 1989 *MUFON UFO Journal*—prior to receiving PT's call—said his research had found nothing to question the "legitimacy" of the MJ-12 papers.

Others have earlier pointed out another suspicious flaw in the alleged Truman memo to Forrestal. This is the fact that the numerical portion of the date—"24, 1947"—was typed using *a different machine* from the one used to type "September."

The logical explanation for this flaw is that the counterfeiter used an old-vintage machine to make it appear that the memo was written in 1947. But the machine's numerical keys were inoperative, forcing the counterfeiter to type the numerical part of the date on a different machine and paste it in. If this were an authentic Truman memo, it would indicate that the President's secretary did not have access to a fully operable typewriter—which is highly unlikely.

Friedman and Moore visited the library to peruse the Bush collection in 1981–1982, prompted by a 1950 memorandum written by Wilbert B. Smith, a Canadian engineer. Smith's memo claimed that the U.S. government was conducting a highly classified investigation into "flying saucers," directed by Bush.

In Moore's paper, presented at a MUFON conference in early July 1982, he reported that he and Friedman had "spent considerable time in

Washington, D.C., over the past year locating and researching dusty files and records. . . ." This enabled him to report that Vannevar Bush and Defense Secretary Forrestal had met with President Truman on September 24, 1947—the date of the MJ-12 memo—after Bush had agreed to head the Pentagon's new research and development board.

A third document made public by Moore, Shandera, and Friedman in the spring of 1987 was what purported to be a "Top Secret" memo from President Eisenhower's special assistant, Robert Cutler, to USAF chief-of-staff Gen. Nathan Twining. The memo, dated July 14, 1954, informed Twining of a slight change of plans for a White House meeting of the "NSC [National Security Council]/MJ-12 Special Studies Project" to be held on July 16.

Moore and Shandera said they found the unsigned carbon copy when they visited the National Archives in mid-1985. As Shandera explained to me, because the memo was found in the National Archives it seemed to officially confirm the existence of MJ-12. However, the Cutler memo lacked a registration number, which all other Top Secret documents in the same files had. Nevertheless, Friedman claimed the memo was authentic because it concluded with "your concurrence in the above change of arrangements is assumed"—almost identical language to that used by Cutler in an earlier memo to Twining, dated July 13, *1953*. Friedman and Moore had found this authentic memo in 1981 in the collection of Twining's papers at the Library of Congress.

Curiously, the MJ-12 Cutler memo was found in recently declassified USAF intelligence material—an unlikely place for a carbon copy seemingly intended for White House files. Also, it had been folded as if it had been carried in the breast pocket of a man's suit. Subsequent investigation by the National Archives revealed that Cutler could not possibly have written the letter because he was out of the country on July 14, 1954. This and other questionable aspects of the document were detailed by a National Archives official in a three-page memorandum.

Did Twining attend an NSC meeting at the White House, as instructed by the MJ-12 Cutler memo? When I checked Twining's official log for July 16, 1954, it showed many appointments but no NSC briefing. When I pointed out this discrepancy to Friedman, he argued that the White House MJ-12 meeting was so secret that it would not be listed in Twining's official log.

If Friedman's logic were valid, then Twining's official log ought not show him attending the "Extraordinary Meeting of the National Security Council" referred to in the authentic Cutler memo of July 13, 1953. Cutler's memo explained that "special security precautions" should be taken "to maintain absolute secrecy regarding participation" in the NSC meeting. For example, Cutler explained that Twining was to enter the White House grounds via a special entrance and his Pentagon limousine should not remain parked near the White House. No such security precautions were prescribed in the MJ-12 Cutler memo.

When I checked Twining's official log in the Library of Congress it *did* show that Twining attended the very secretive NSC conference in 1953. His log showed: "National Security Council at White House all day"—demolishing Friedman's claim. By a curious coincidence, this secret July 16, 1953, NSC meeting was held one year to the day of the alleged MJ-12 NSC meeting.

Ironically, in the introduction to a paper on crashed-saucer claims authored by Moore and Friedman, presented at the 1981 MUFON conference, they quoted Albert Einstein as follows: "The right to search for the truth implies also a duty; one must not conceal any part of what one has recognized to be the truth." This recalls the admonition by French philosopher Charles Peguy: "He who does not bellow the truth when he knows the truth makes himself the accomplice of liars and forgers."

EDITOR'S NOTE

William L. Moore was informed of the investigation and conclusions reported above. In a letter(October 16, 1989), Moore acknowledged that the document examiner referred to as PT had indeed made his (hoax) findings available "some time ago" and "we have not yet published them." But, he said, PT was only one of four document examiners he and his colleagues had consulted and claimed the opinions of the four about the issues involved with the Truman document are "mixed." He did not name the other examiners. Moore said that a report would be published soon.

10.
TOP-SECRET BALLOON PROJECT LOOMS OVER TV MOVIE ON ROSWELL INCIDENT

C. Eugene Emery, Jr.

At the conclusion of *Roswell,* the made-for-cable-television movie about the alleged crash of an extraterrestrial spacecraft in New Mexico in 1947, an enigmatic UFO investigator named Townsend confronts Major Jesse Marcel, who helped recover the debris and is trying to expose the U.S. government's efforts to hide the truth.

"You have nothing, just a lot of old memories and secondhand recollections," Townsend scolds Marcel, who is battered by years of ridicule for identifying the debris as extraterrestrial when the government insisted that the material was from a weather balloon. "Nobody is going to take you seriously, not without proof, not without hard evidence."

Somebody should have made that speech to Showtime, the cable outfit that bankrolled the movie, and its executive producer, Paul Davids, who says he saw a UFO in 1987. Perhaps they would have thought twice about creating this strange film, which premiered July 31 and promised "startling new information about an historic UFO incident."

There was nothing new here.

Roswell (the incident) has been short on facts and long on both speculation and secondhand memories since its origins in 1947.

Roswell (the movie) seems contrived to gloss over facts that contradict a UFO explanation and convince doubters that an alien spacecraft did indeed crash in a remote field. The film shows alien hieroglyphics, metal that can't be broken, and foil that de-wrinkles after it is folded.

But *Roswell* doesn't stop there. By the end of the film, viewers are

---

This article originally appeared in the *Skeptical Inquirer* 19 (January/February 1995). Reprinted with permission.

told that four extraterrestrials were recovered near the crash site and that one of them lived long enough to send a telepathic message to U.S. Secretary of War James Forrestal.

Look closely and you can see that *Roswell*—like the alleged disinformation campaign that supposedly kept the UFO crash secret—is carefully calculated to neutralize the criticisms of the skeptics.

Do the critics carp that only a small amount of debris was recovered? The solution for the filmmakers is to show a huge area littered with truckloads of shiny, silvery alien stuff.

Do the critics insist that the recovered material came from the weather balloon? Then show a weather balloon being launched from the center of town and have the characters insisting that it couldn't have been a weather balloon because everyone knows what a weather balloon looks like.

Some of the holes in *Roswell* seem large enough to fly a Mother Ship through.

The local undertaker says he was called by someone at the military base looking for child-sized caskets and information on how to preserve bodies. Hadn't anyone on the base ever heard of refrigeration? Wasn't anyone on the base smart enough to bag the bodies and put them on ice?

Drop a dead alien on my doorstep and I'll want the autopsy done at a top-notch medical facility. But in *Roswell,* the autopsy is performed at the base, in a room crammed with (by my count) twenty-eight people, sixteen of whom weren't even wearing masks to guard against infection by an extraterrestrial virus or bacterium.

During the autopsy, one of the doctors announces that the creature has no digestive system. Perhaps the ETs are solar powered.

And in the scene where the surviving alien telepathically communicates with Forrestal, the creature is shown lying on black sheets (how many hospitals have *those*?) and his attendants are, once again, wearing no masks to protect either themselves or the alien from illnesses to which they might have no immunity.

* * *

Coincidentally, the movie came out just as new evidence was surfacing that provided a convincing explanation for why the U.S. government has been cagey about the Roswell case.

The debris found at the ranch of Mac Brazel has been identified as a six-sided balloon-borne radar reflector launched as part of a secret military effort code named Project Mogul. The "hieroglyphics" reportedly found on the debris were apparently printed on patterned seamstress tape used in making the World War II–era radar reflectors.

Credit for finding the connection to Project Mogul goes to Robert G. Todd of Ardmore, Pennsylvania, whose Freedom of Information Act request produced the now-declassified documents about the project. Project Mogul consisted of a series of experiments to use high-altitude balloons to carry instruments aloft in an attempt to detect acoustic signatures perhaps propagated along the boundary between the troposphere and stratosphere from Soviet nuclear tests. The radar reflectors let the military keep track of the balloons. Another UFO researcher, Karl Pflock has also documented the link.

In September, the air force issued a report of its own investigation into the Roswell case, confirming, and providing more details about, Project Mogul. It even identified the debris as that of a specific, never-found Project Mogul balloon flight package launched in June 1947.*

Project Mogul scientist Charles B. Moore, now at the New Mexico Institute of Mining and Technology, but then of New York University, did not realize the connection until Todd gave him a copy of Brazel's original description of the debris, published in the July 9, 1947, *Roswell Daily Record.* The paper said "the tinfoil, paper, tape and sticks made a bundle about three feet long and seven or eight inches thick" and rubber strips "made a bundle about 18 or 20 inches long and about 8 inches thick. In all, [Brazel] estimated, the entire lot would have weighed maybe five pounds."

Todd said Moore "immediately saw the significance of it because nobody had shown him the text (of the original description) before. It was obvious what it was."

The description in the *Roswell Daily Record* noted that "considerable scotch tape and some tape with flowers printed on it had been used in the construction." Moore suggested that the reflectors may have contained patterned tape because such reflectors were sometimes made in garment lofts in Manhattan.

* * *

Powerful documentaries often conclude with an epilogue designed to clinch the argument a filmmaker is trying to make. *Roswell* ends with several assertions that sound compelling only until you think about them:

• Since Marcel died in 1986, "over 350 witnesses to the event have agreed to talk." But even if you believe the movie version of the facts, too many of those witnesses were never in a position to have firsthand knowledge of the alleged events. The town undertaker never saw the bodies; he

---

*See "For the Record," *Skeptical Inquirer* 19 (January 1995): 41.

simply claims to know someone who did. Nobody saw the bodies of the aliens as they were being moved around, although they know somebody who might have. The radio newsman Frank Joyce is shown describing a government interrogation of Brazel that Brazel never acknowledges and that Joyce never witnessed.

• "Officially, the military denies that UFOs are extraterrestrial," which suggests that because the government officially denies it, it must be true. But just because the U.S. government might officially deny the existence of the tooth fairy doesn't automatically mean an ill-mannered dwarf is stalking our children and stealing their newly lost teeth.

• "The U.S. government has canceled its project to search for signs of extraterrestrial intelligence." I guess this is supposed to imply they canceled the search because they found something. That's news to Carl Sagan and others who are actively involved in the search. The fact is, a budget-conscious Congress cut off money for the search, which continues privately.

• "One in ten Americans claims to have seen a UFO." Guess that means Showtime's next documentary will show us the secret life of Santa Claus.

Another epilogue would have been more accurate: "Although 350 claim to be witnesses to the events described here, despite truckloads of debris with extraordinary physical characteristics, despite the windfall of four (count 'em, *four*) alien bodies available for examination, no reputable scientist has come forward to share some of the wealth of scientific knowledge such remains would undoubtedly produce. Nor is there any evidence that such knowledge has done anything to fuel the breakthroughs in physics, chemistry, engineering, or biology that have occurred in the intervening 47 years."

11. AIR FORCE REPORT ON THE ROSWELL INCIDENT

Richard L. Weaver, Col. USAF

The following "Memorandum for Correspondents" was issued by the Secretary of the Air Force September 8, 1994, accompanying a twenty-three-page report of a 1994 U.S. Air Force investigation into the so-called Roswell incident, a claim of a crashed flying saucer in New Mexico in 1947. Following the memorandum we print a substantial portion of the Report of Air Force Research Regarding the "Roswell Incident," *including its Executive Summary, the key section, "What the 'Roswell Incident' Was," and the Conclusion. The report provides abundant evidence identifying the debris recovered on a New Mexico ranch as that from a balloon, a radar-tracking reflector, and an instrument package from "Project Mogul." Project Mogul was then a Top Secret field experiment being carried out at Alamogordo, New Mexico, by New York University scientists to study the possibility of detecting low-frequency pressure waves in the atmosphere from Soviet nuclear weapons tests.*

MEMORANDUM FOR CORRESPONDENTS

No. 255-M, Sept. 8, 1994

Secretary of the Air Force Sheila E. Widnall today announced the completion of an Air Force study to locate records that would explain an alleged 1947 UFO incident. Pro-UFO researchers claim an extraterrestrial

---

This article originally appeared in the *Skeptical Inquirer* 19 (January/February 1995). Reprinted with permission.

spacecraft and its alien occupants were recovered near Roswell, N.M., in July 1947 and the fact was kept from the public.

At the request of Congressman Steven H. Schiff (R-N.Mex.), the General Accounting Office in February 1994 initiated an audit to locate all records related to the Roswell incident and to determine if such records were properly handled. The GAO audit entitled "Records Management Procedures Dealing With Weather Balloon, Unknown Aircraft, and Similar Crash Incidents" is not yet complete.

The GAO audit involved a number of government agencies but focused on the Air Force. In support of the GAO effort, the Air Force initiated a systematic search of current Air Force offices as well as numerous archives and records centers that might help explain the incident. Air Force officials also interviewed a number of persons who might have had knowledge of the events. Prior to the interviews Secretary Widnall released those persons from any previous security obligations that might have restricted their statements.

The Air Force research did not locate or develop any information that the "Roswell Incident" was a UFO event nor was there any indication of a "cover-up" by the Air Force. Information obtained through exhaustive records searches and interviews indicated the material recovered near Roswell was consistent with a balloon device of the type used in a then-classified project. No records indicated or even hinted at the recovery of "alien" bodies or extraterrestrial materials.

All documentation related to this case is now declassified and the information is in the public domain. All documentation has been turned over to the Air Force Historian. The Air Force report without attachments may be obtained by contacting Major Thurston, Air Force Public Affairs, (703) 695-0640. The report with all 33 attachments is available for review in the Pentagon Library in Room 1A518.

REPORT OF AIR FORCE RESEARCH REGARDING THE "ROSWELL INCIDENT"

Executive Summary

The "Roswell Incident" refers to an event that supposedly happened in July, 1947, wherein the Army Air Forces (AAF) allegedly recovered remains of a crashed "flying disc" near Roswell, New Mexico. In February, 1994, the General Accounting Office (GAO), acting on the request of a New Mexico Congressman, initiated an audit to attempt to locate records of such an incident and to determine if records regarding it were properly handled. Although the GAO effort was to look at a number of government agencies, the apparent focus was on the Air Force. SAF/AAZ [Secretary of the Air Force/Security and Special Program Oversight], as the Central Point of Contact for the GAO in this matter, initiated a systematic search of current Air Force offices as well as numerous archives and records centers that might help explain this matter. Research revealed that the "Roswell Incident" was not even considered a UFO event until the 1978–1980 time frame. Prior to that, the incident was dismissed because the AAF originally identified the debris recovered as being that of a weather balloon. Subsequently, various authors wrote a number of books claiming that, not only was debris from an alien spacecraft recovered, but also the bodies of the craft's alien occupants. These claims continue to evolve today and the Air Force is now routinely accused of engaging in a "cover-up" of this supposed event.

The research located no records at existing Air Force offices that indicated any "cover-up" by the USAF or any indication of such a recovery. Consequently, efforts were intensified by Air Force researchers at numerous locations where records for the period in question were stored. The records reviewed did not reveal any increase in operations, security, or any other activity in July, 1947, that indicated any such unusual event may have occurred. Records were located and thoroughly explored concerning a then-TOP SECRET balloon project, designed to attempt to monitor Soviet nuclear tests, known as Project Mogul. Additionally, several surviving project personnel were located and interviewed, as was the only surviving person who recovered debris from the original Roswell site in 1947, and the former officer who initially identified the wreckage as a balloon. Comparison of all information developed or obtained indicated that the material recovered near Roswell was consistent with a balloon device and most likely from one of the Mogul balloons that had not been previously recovered. Air Force research efforts did not disclose any records of the recovery of any "alien" bodies or extraterrestrial materials.

Introduction

. . . Even though Air Force research originally started in January, 1994, the first official Air Force-wide tasking was directed by the March 1, 1994, memorandum from SAF/AA, (Atch 5), and was addressed to those current Air Staff elements that would be the likely repository for any records, particularly if there was anything of an extraordinary nature involved. This meant that the search was not limited to unclassified materials, but also would include records of the highest classification and compartmentation. . . .

What the Roswell Incident Was Not

Before discussing specific positive results that these efforts revealed, it is first appropriate to discuss those things, as indicated by information available to the Air Force, that the "Roswell Incident" was not:

An Airplane Crash . . .

A Missile Crash . . .

A Nuclear Accident . . .

An Extraterrestrial Craft.

The Air Force research found absolutely no indication that what happened near Roswell in 1947 involved any type of extraterrestrial spacecraft. This, of course, is the crux of this entire matter. "Pro-UFO" persons who obtain a copy of this response, at this point, most probably begin the "cover-up is still on" claims. Nevertheless, the research indicated absolutely no evidence *of any kind* that a spaceship crashed near Roswell or that any alien occupants were recovered therefrom, in some secret military operation or otherwise. This does not mean, however, that the early Air Force was not concerned about UFOs. However, in the early days, "UFO" meant Unidentified Flying Object, which literally translated as some object in the air that was not readily identifiable. It did not mean, as the term has evolved in today's language, to equate to alien spaceships. Records from the period reviewed by Air Force researchers as well as those cited by the authors mentioned before, do indicate that the USAF *was* seriously concerned about the inability to adequately identify unknown flying objects reported in American airspace. All the records, however, indicated that the focus of concern was not on aliens, hostile or otherwise, but on the Soviet Union. Many documents from that period speak to the possibility of developmental secret Soviet aircraft overflying U.S. airspace. This, of course, was of major concern to the fledgling USAF, whose job it was to protect these same skies.

The research revealed only one official AAF document that indicated that there was any activity of any type that pertained to UFOs and Roswell in July, 1947. This was a small section of the July Historical Report for the 509th Bomb Group and Roswell AAF that stated: "*The Office of Public Information* was quite busy during the month answering inquiries on the 'flying disc,' which was reported to be in possession of the 509th Bomb Group. The object turned out to be a radar tracking balloon" (included with Atch 11). Additionally, this history showed that the 509th Commander, Colonel Blanchard, went on leave on July 8, 1947, which would be a somewhat unusual maneuver for a person involved in the supposed first ever recovery of extraterrestrial materials. (Detractors claim Blanchard did this as a ploy to elude the press and go to the scene to direct the recovery operations.) The history and the morning reports also showed that the subsequent activities at Roswell during the month were mostly mundane and not indicative of any unusual high level activity, expenditure of manpower, resources, or security.

Likewise, the researchers found no indication of heightened activity anywhere else in the military hierarchy in the July, 1947, message traffic or orders (to include classified traffic). There were no indications and warnings, notice of alerts, or a higher tempo of operational activity reported that would be logically generated if an alien craft, whose intentions were unknown, entered US territory. . . .

What the "Roswell Incident" Was

As previously discussed, what was originally reported to have been recovered was a balloon of some sort, usually described as a "weather balloon," although the majority of the wreckage that was ultimately displayed by General Ramey and Major Marcel in the famous photos (Atch 16) in Ft. Worth, was that of a radar target normally suspended from balloons. This radar target, discussed in more detail later, was certainly consistent with the description of a July 9 newspaper article which discussed "tinfoil, paper, tape, and sticks." Additionally, the description of the "flying disc" was consistent with a document routinely used by most pro-UFO writers to indicate a conspiracy in progress—the telegram from the Dallas FBI office of July 8, 1947. This document quoted in part states: ". . . The disc is hexagonal in shape and was suspended from a balloon by a cable, which balloon was approximately twenty feet in diameter. . . . The object found resembles a high altitude weather balloon with a radar reflector. . . . Disc and balloon being transported. . . ."

Similarly, while conducting the popular literature review, one of the documents reviewed was a paper entitled "The Roswell Events," edited by Fred Whiting and sponsored by the Fund for UFO Research (FUFOR). Although it was not the original intention to comment on what commercial authors interpreted or claimed that other persons supposedly said, this particular document was different because it contained actual copies of apparently authentic sworn affidavits received from a number of persons who claimed to have some knowledge of the Roswell event. Although many of the persons who provided these affidavits to the FUFOR researchers also expressed opinions that they thought there was something extraterrestrial about this incident, a number of them actually described materials that sounded suspiciously like wreckage from balloons. These included the following:

Jesse A. Marcel, M.D. (son of the late Major Jesse Marcel; 11 years old at the time of the incident). Affidavit dated May 6, 1991. ". . . There were three categories of debris: a thick, foil-like metallic gray substance; a brittle, brownish-black plastic-like material, like Bakelite; and there were fragments of what appeared to be I-beams. On the inner surface of the I-beam, there appeared to be a type of writing. This writing was a purple-violet hue, and it had an embossed appearance. The figures were composed of curved, geometric shapes. It had no resemblance to Russian, Japanese, or any other foreign language. It resembled hieroglyphics, but it had no animal-like characters. . . ."

Bessie Brazel Schreiber (daughter of W. W. Brazel; fourteen years old at the time of the incident). Affidavit dated September 22, 1993. ". . . The debris looked like pieces of a large balloon which had burst. The pieces were small, the largest I remember measuring about the same as the diameter of a basketball. Most of it was a kind of double-sided material, foil-like on one side and rubber-like on the other. Both sides were grayish silver in color, the foil more silvery than the rubber. Sticks, like kite sticks, were attached to some of the pieces with a whitish tape. The tape was about two or three inches wide and had flowerlike designs on it. The 'flowers' were faint, a variety of pastel colors, and reminded me of Japanese paintings in which the flowers are not all connected. I do not recall any other types of material or markings, nor do I remember seeing gouges in the ground or any other signs that anything may have hit the ground hard. The foil-rubber material could not be torn like ordinary aluminum foil can be torn. . . ."

In addition to those persons above still living who claim to have seen or examined the original material found on the Brazel Ranch, there is one additional person who was universally acknowledged to have been involved in its recovery, Sheridan Cavitt, Lt. Col., USAF (Ret). Cavitt is credited in all claims of having accompanied Major Marcel to the ranch to recover the debris, sometimes along with his Counter Intelligence Corps (CIC) subordinate, William Rickett, who, like Marcel, is deceased. Although there does not appear to be much dispute that Cavitt was involved in the material recovery, other claims about him prevail in the popular literature. He is sometimes portrayed as a closed-mouth (or sometimes even sinister) conspirator who was one of the early individuals who kept the "secret of Roswell" from getting out. Other things about him have been alleged, including the claim that he wrote a report of the incident at the time that has never surfaced.

Since Lt. Col. Cavitt, who had firsthand knowledge, was still alive, a decision was made to interview him and get a signed sworn statement from him about his version of the events. Prior to the interview, the Secretary of the Air Force provided him with a written authorization and waiver to discuss classified information with the interviewer and release him from any security oath he may have taken. Subsequently, Cavitt was interviewed on May 24, 1994, at his home. Cavitt provided a signed, sworn statement (Atch 17) of his recollections in this matter. He also consented to having the interview tape-recorded. A transcript of that recording is at Atch 18. In this interview, Cavitt related that he had been contacted on numerous occasions by UFO researchers and had willingly talked with many of them; however, he felt that he had oftentimes been misrepresented or had his comments taken out of context so that their true meaning was changed. He stated unequivocally, however, that the material he recovered consisted of a reflective sort of material like aluminum foil and some thin, bamboo-like sticks. He thought at the time, and continued to do so today, that what he found was a weather balloon and has told other private researchers that. He also remembered finding a small "black box" type of instrument, which he thought at the time was probably a radiosonde. Lt. Col. Cavitt also reviewed the famous Ramey/Marcel photographs (Atch 16) of the wreckage taken to Ft. Worth (often claimed by UFO researchers to have been switched and the remnants of a balloon substituted for it) and he identified the materials depicted in those photos as consistent with the materials that he recovered from the ranch. Lt. Col. Cavitt also stated that he had never taken any oath or signed any agreement not to talk about this incident and had never been threatened by anyone in the government because of it. He did not even know the "incident" was claimed to be anything unusual until he was interviewed in the early 1980s.

Similarly, Irving Newton, Major, USAF (Ret) was located and interviewed. Newton was a weather officer assigned to Fort Worth, who was on duty when the Roswell debris was sent there in July, 1947. He was told that he was to report to General Ramey's office to view the material. In a signed, sworn statement (Atch 30) Newton related that ". . . I walked into the General's office where this supposed flying saucer was lying all over the floor. As soon as I saw it, I giggled and asked if that was the flying saucer. . . . I told them that this was a balloon and a RAWIN target. . . ." Newton also stated that ". . . while I was examining the debris, Major Marcel was picking up pieces of the target sticks and trying to convince me that some notations on the sticks were alien writings. There were figures on the sticks, lavender or pink in color, appeared to be weather faded markings, with no rhyme or reason [*sic*]. He did not convince me that these were alien writings." Newton concluded his statement by relating that ". . . during the ensuing years I have been interviewed by many authors, I have been quoted and misquoted. The facts remain as indicated above. I was not influenced during the original interview, nor today, to provide anything but what I know to be true, that is, the material I saw in General Ramey's office was the remains of a balloon and a RAWIN target."

Balloon Research

The original tasking from GAO noted that the search for information included "weather balloons." Comments about balloons and safety reports have already been made; however, the SAF/AAZ research efforts also focused on reviewing historical records involving balloons, since, among other reasons, that was what was officially claimed by the AAF to have been found and recovered in 1947.

As early as February 28, 1994, the AAZD research team found references to balloon tests taking place at Alamogordo AAF (now Holloman AFB) and White Sands during June and July 1947, testing "constant level balloons" and a New York University (NYU)/Watson Labs effort that used ". . . meteorological devices . . . suspected for detecting shock waves generated by Soviet nuclear explosions"—a possible indication of a cover story associated with the NYU balloon project. Subsequently, a 1946 HQ AMC memorandum was surfaced, describing the constant altitude balloon project, and specified that the scientific data be classified TOP SECRET Priority 1A. Its name was Project Mogul (Atch 19).

Project Mogul was a then-sensitive, classified project, whose purpose was to determine the state of Soviet nuclear weapons research. This was the early Cold War period and there was serious concern within the U.S. government about the Soviets developing a weaponized atomic device. Because the Soviet Union's borders were closed, the U.S. government sought to develop a long range nuclear explosion detection capability. Long range, balloon-borne, low frequency acoustic detection was posed to General Spaatz in 1945 by Dr. Maurice Ewing of Columbia University as a potential solution (atmospheric ducting of low frequency pressure waves had been studied as early as 1900).

As part of the research into this matter, AAZD personnel located and obtained the original study papers and reports of the New York University project. Their efforts also revealed that some of the individuals involved in Project Mogul were still living. These persons included the NYU constant altitude balloon Director of Research, Dr. Athelstan F. Spilhaus; the Project Engineer, Professor Charles B. Moore; and the military Project Officer, Colonel Albert C. Trakowski.

All of these persons were subsequently interviewed and signed sworn statements about their activities. A copy of these statements are appended at Atch 20-22. Additionally, transcripts of the interview with Moore and Trakowski are also included (equipment malfunctioned during the interview of Spilhaus) (Atch 23-24). These interviews confirmed that Project Mogul was a compartmented, sensitive effort. The NYU group was responsible for developing constant level balloons and telemetering equipment that would remain at specified altitudes (within the acoustic duct) while a group from Columbia was to develop acoustic sensors. Doctor Spilhaus, Professor Moore, and certain others of the group were aware of the actual purpose of the project, but they did not know of the project nickname at the time. They handled casual inquiries and/or scientific inquiries/papers in terms of "unclassified meteorological or balloon research." Newly hired employees were not made aware that there was anything special or classified about their work; they were told only that their work dealt with meteorological equipment.

An advance ground team, led by Albert P. Crary, preceded the NYU group to Alamogordo AAF, New Mexico, setting up ground sensors and obtaining facilities for the NYU group. Upon their arrival, Professor Moore and his team experimented with various configurations of neoprene balloons; development of balloon "trains" (see illustration, Atch 25); automatic ballast systems; and use of Naval sonobuoys (as the Watson Lab acoustical sensors had not yet arrived). They also launched what they called "service flights." These "service flights" were not logged or fully accounted for in the published Technical Reports generated as a result of the contract between NYU and Watson Labs. According to Professor Moore, the "service flights" were composed of balloons, radar reflectors, and payloads specifically designed to test acoustic sensors (both early sonobuoys and the later Watson Labs devices). The "payload equipment" was expendable and some carried no "REWARD" or "RETURN TO . . ." tags because there was to be no association between these flights and the logged constant altitude flights which were fully acknowledged. The NYU balloon flights were listed sequentially in their reports (i.e., A, B, 1, 5, 6, 7, 8, 10 . . .) yet gaps existed for Flights 2-4 and Flight 9. The interview with Professor Moore indicated that these gaps were the unlogged "service flights."

Professor Moore, the on-scene Project Engineer, gave detailed information concerning his team's efforts. He recalled that radar targets were used for tracking balloons because they did not have all the necessary equipment when they first arrived in New Mexico. Some of the early developmental radar targets were manufactured by a toy or novelty company. These targets were made up of aluminum "foil" or foil-backed paper, balsa wood beams that were coated in an "Elmer's-type" glue to enhance their durability, acetate and/or cloth reinforcing tape, single strand and braided nylon twine, brass eyelets and swivels to form a multi-faced reflector somewhat similar in construction to a box kite (see photographs, Atch 26). Some of these targets were also assembled with purplish-pink tape with symbols on it (see drawing by Moore with Atch 21).

According to the log summary (Atch 27) of the NYU group, Flight A through Flight 7 (November 20, 1946 to July 2, 1947) were made with neoprene meteorological balloons (as opposed to the later flights made with polyethylene balloons). Professor Moore stated that the neoprene balloons were susceptible to degradation in the sunlight, turning from a milky white to a dark brown. He described finding remains of balloon trains with reflectors and payloads that had landed in the desert: the ruptured and shredded neoprene would "almost look like dark gray or black flakes or ashes after exposure to the sun for only a few days. The plasticizers and antioxidants in the neoprene would emit a peculiar acrid odor and the balloon material and radar target material would be scattered after returning to earth depending on the surface winds." Upon review of the local newspaper photographs from General Ramey's press conference in 1947 and descriptions in popular books by individuals who supposedly handled the debris recovered on the ranch, Professor Moore opined that the material was most likely the shredded remains of a multi-neoprene balloon train with multiple radar reflectors. The material and a "black box," described by Cavitt, was, in Moore's scientific opinion, most probably from Flight 4, a "service flight" that included a cylindrical metal sonobuoy and portions of a weather instrument housed in a box, which was unlike typical weather radiosondes which were made of cardboard. Additionally, a copy of a professional journal, maintained at the time by A. P. Crary, provided to the Air Force by his widow, showed that Flight 4 was launched on June 4, 1947, but was not recovered by the NYU group. It is very probable that this TOP SECRET project balloon train (Flight 4), made up of unclassified components, came to rest some miles northwest of Roswell, N.M., became shredded in the surface winds and was ultimately found by the rancher, Brazel, ten days later. This possibility was supported by the observations of Lt. Col. Cavitt (Atch 17-18), the only living eyewitness to the actual debris field and the material found. Lt. Col. Cavitt described a small area of debris which appeared "to resemble bamboo type square sticks one-quarter to one-half inch square, that were very light, as well as some sort of metallic reflecting material that was also very light . . . I remember recognizing this material as being consistent with a weather balloon."

Concerning the initial announcement, "RAAF Captures Flying Disc," research failed to locate any documented evidence as to why that statement was made. However, on July10, 1947, following the Ramey press conference, the *Alamogordo News* published an article with photographs demonstrating multiple balloons and targets at the same location as the NYU group operated from at Alamogordo AAF. Professor Moore expressed surprise at seeing this since his was the only balloon test group in the area. He stated, "It appears that there was some type of umbrella cover story to protect our work with Mogul." Although the Air Force did not find documented evidence that Gen. Ramey was directed to espouse a weather balloon in his press conference, he may have done so either because he was aware of Project Mogul and was trying to deflect interest from it or because he readily perceived the material to be a weather balloon based on the identification from his weather officer, Irving Newton. In either case, the materials recovered by the AAF in July 1947, were not readily recognizable as anything special (only the purpose was special) and the recovered debris itself was unclassified. Additionally, the press dropped its interest in the matter as quickly as they had jumped on it. Hence, there would be no particular reason to further document what quickly became a "non-event."

The interview with Colonel Trakowski (Atch 23–24) also proved valuable information. Trakowski provided specific details on Project Mogul and described how the security for the program was set up, as he was formerly the TOP SECRET Control Officer for the program. He further related that many of the original radar targets that were produced around the end of World War II were fabricated by toy or novelty companies using a purplish-pink tape with flower and heart symbols on it. Trakowski also recounted a conversation that he had had with his friend, and superior military officer in his chain of command, Colonel Marcus Duffy, in July, 1947. Duffy formerly had Trakowski's position on Mogul, but had subsequently been transferred to Wright Field. He stated: "Colonel Duffy called me on the telephone from Wright Field and gave me a story about a fellow that had come in from New Mexico, woke him up in the middle of the night or some such thing with a handful of debris, and wanted him, Colonel Duffy, to identify it.

". . . He just said, 'It sure looks like some of the stuff you've been launching at Alamogordo' and he described it, and I said, 'Yes, I think it is.' Certainly Colonel Duffy knew enough about radar targets, radiosondes, balloon-borne weather devices. He was intimately familiar with all that apparatus."

Attempts were made to locate Colonel Duffy, but it was ascertained that he had died. His widow explained that, although he had amassed a large amount of personal papers relating to his Air Force activities, she had recently disposed of these items. Likewise, it was learned that A. P. Crary was also deceased; however his surviving spouse had a number of his papers from his balloon-testing days, including his professional journal from the period in question. She provided the Air Force researchers with this material. It is discussed in more detail within Atch 32. Overall, it helps fill in gaps of the Mogul story.

During the period the Air Force conducted this research, it was discovered that several others had also discovered the possibility that the "Roswell Incident" may have been generated by the recovery of a Project Mogul balloon device. These persons included Professor Charles B. Moore, Robert Todd, and coincidentally, Karl Pflock, a researcher who is married to a staffer who works for Congressman Schiff. Some of these persons provided suggestions as to where documentation might be located in various archives, histories, and libraries. A review of Freedom of Information Act (FOIA) requests revealed that Robert Todd, particularly, had become aware of Project Mogul several years ago and had doggedly obtained from the Air Force, through the FOIA, a large amount of material pertaining to it, long before the AAZD researchers independently seized on the same possibility.

Most interesting, as this report was being written, Pflock published his own report of this matter under the auspices of FUFOR, entitled "Roswell in Perspective" (1994). Pflock concluded from his research that the Brazel Ranch debris originally reported as a "flying disc" was probably debris from a Mogul balloon, [but that] there was a simultaneous incident that occurred not far away that caused an alien craft to crash and that the AAF subsequently recovered three alien bodies therefrom. Air Force research did not locate any information to corroborate that this incredible coincidence occurred, however.

In order to provide a more detailed discussion of the specifics of Project Mogul and how it appeared to be directly responsible for the "Roswell Incident," a SAF/AAZD researcher prepared a more detailed discussion on the balloon project which is appended to this report as Atch 32.

Other Research

In the attempt to develop additional information that could help explain this matter, a number of other steps were taken. First, assistance was requested from various museums and other archives (Atch 28) to obtain information and/or examples of the actual balloons and radar targets used in connection with Project Mogul and to correlate them with the various descriptions of wreckage and materials recovered. The blueprints for the "Pilot Balloon Target ML307C/AP Assembly" (generically, the radar target assembly) were located at the Army Signal Corps Museum at Fort Monmouth and obtained. A copy is appended as Atch 29. This blueprint provides the specification for the foil material, tape, wood, eyelets, and string used and the assembly instructions thereto. An actual device was also obtained for study with the assistance of Professor Moore. (The example actually procured was a 1953-manufactured model "C" as compared to the Model B which was in use in 1947. Professor Moore related that the differences were minor.) An examination of this device revealed it to be simply made of aluminum-colored foil-like material over a stronger paper-like material, attached to balsa wood sticks, affixed with tape, glue, and twine. When opened, the device appears as depicted in Atch 31 (contemporary photo) and Atch 25 (1947 photo, in a "balloon train"). When folded, the device is in a series of triangles, the largest being four feet by two feet ten inches. The smallest triangle section measures two feet by two feet ten inches. (Compare with descriptions provided by Lt. Col. Cavitt and others, as well as photos of wreckage.)

Additionally, the researchers obtained from the Archives of the University of Texas–Arlington (UTA) a set of original (i.e., first generation) prints of the photographs taken at the time by the *Fort Worth Star-Telegram,* that depicted Ramey and Marcel with the wreckage. A close review of these photos (and a set of first generation negatives also subsequently obtained from UTA) revealed several interesting observations. First, although in some of the literature cited above, Marcel allegedly stated that he had his photo taken with the "real" UFO wreckage and then it was subsequently removed and the weather balloon wreckage substituted for it, a comparison shows that the same wreckage appeared in the photos of Marcel and Ramey. The photos also depicted that this material was lying on what appeared to be some sort of wrapping paper (consistent with affidavit excerpt of crew chief Porter, above). It was also noted that in the two photos of Ramey he had a piece of paper in his hand. In one, it was folded over so nothing could be seen. In the second, however, there appears to be text printed on the paper. In an attempt to read this text to determine if it could shed any further light on locating documents relating to this matter, the photo was sent to a national level organization for digitizing and subsequent photo interpretation and analysis. This organization was also asked to scrutinize the digitized photos for any indication of the flowered tape (or "hieroglyphics, depending on the point of view) that were reputed to be visible to some of the persons who observed the wreckage prior to its getting to Fort Worth. This organization reported on July 20, 1994, that even after digitizing, the photos were of insufficient quality to visualize either of the details sought for analysis. This organization was able to obtain measurements from the "sticks" visible in the debris after it was ascertained by an interview of the original photographer what kind of camera he used. The results of this process are provided in Atch 33, along with a reference diagram and the photo from which the measurements were made. All these measurements are compatible with the wooden materials used in the radar target previously described.

Conclusion

The Air Force research did not locate or develop any information that the "Roswell Incident" was a UFO event. All available official materials, although they do not directly address Roswell per se, indicate that the most likely source of the wreckage recovered from the Brazel Ranch was from one of the Project Mogul balloon trains. Although that project was TOP SECRET at the time, there was also no specific indication found to indicate that an official preplanned cover story was in place to explain an event such as that which ultimately happened. It appears that the identification of the wreckage as being part of a weather balloon device, as reported in the newspapers at the time, was based on the fact that there was no physical difference in the radar targets and the neoprene balloons (other than the numbers and configuration) between Mogul balloons and normal weather balloons. Additionally, it seems that there was overreaction by Colonel Blanchard and Major Marcel, in originally reporting that a "flying disc" had been recovered when, at that time, nobody for sure knew what that term even meant since it had only been in use for a couple of weeks.

Likewise, there was no indication in official records from the period that there was heightened military operational or security activity which should have been generated if this was, in fact, the first recovery of materials and/or persons from another world. The postwar U.S. military (or today's for that matter) did not have the capability to rapidly identify, recover, coordinate, cover up, and quickly minimize public scrutiny of such an event. The claim that they did so without leaving even a little bit of a suspicious paper trail for 47 years is incredible.

It should also be noted here that there was little mentioned in this response about the recovery of the so-called alien bodies. This is for several reasons: First, the recovered wreckage was from a Project Mogul balloon. There were no "alien" passengers therein. Secondly, the pro-UFO groups who espouse the alien bodies theories cannot even agree among themselves as to what, how many, and where, such bodies were supposedly recovered. Additionally, some of these claims have been shown to be hoaxes, even by other UFO researchers. Thirdly, when such claims are made, they are often attributed to people using pseudonyms or who otherwise do not want to be publicly identified, presumably so that some sort of retribution cannot be taken against them (notwithstanding that nobody has been shown to have died, disappeared or otherwise suffered at the hands of the government during the last 47 years). Fourth, many of the persons making the biggest claims of "alien bodies" make their living from the "Roswell Incident." While having a commercial interest in something does not automatically make it suspect, it does raise interesting questions related to authenticity. Such persons should be encouraged to present their evidence (not speculation) directly to the government and provide all pertinent details and evidence to support their claims if honest fact-finding is what is wanted. Lastly, persons who have come forward and provided their names and made claims, may have, in good faith but in the "fog of time," misinterpreted past events. The review of Air Force records did not locate even one piece of evidence to indicate that the Air Force has had any part in an "alien" body recovery operation or continuing coverup.

During the course of this effort, the Air Force has kept in close touch with the GAO and responded to their various queries and requests for assistance. This report was generated as an official response to the GAO, and to document the considerable effort expended by the Air Force on their behalf. It is anticipated that that they will request a copy of this report to help formulate the formal report of their efforts. It is recommended that this document serve as the final Air Force report related to the Roswell matter, for the GAO, or any other inquiries.

12. THE ROSWELL INCIDENT AND PROJECT MOGUL

David E. Thomas

As reported in the January–February 1995 *Skeptical Inquirer*, a September 1994 air force report strongly supported the theory that the "UFO" debris found by rancher Mac Brazel in 1947 northwest of Roswell, New Mexico, was in fact a remnant of a balloon flight launched as part of a top-secret program called Project Mogul. The possible connection between the Roswell Incident and Mogul was first realized by researcher Robert G. Todd, and independently by Karl T. Pflock.

Recently, Charles B. Moore, one of three surviving Project Mogul scientists identified in and interviewed for the air force report, spoke to the New Mexicans for Science and Reason (NMSR) in Albuquerque. He discussed the background of the project, the New York University (NYU) balloon flights, and the Roswell connection. He provided new details that would appear to virtually clinch the idea that the debris Brazel found was indeed from one of the Project Mogul flights that Moore helped launch.

What follows is based on Moore's presentation, his answers to audience questions, subsequent meetings and discussions with him, documents he provided, and a monograph he is preparing on these flights.

Moore, professor emeritus of physics at New Mexico Institute of Mining and Technology in Socorro, was a graduate student working for NYU back in 1947. The Mogul project was so classified and compartmentalized that even Moore didn't know the project's name until Robert Todd informed him of it a couple of years ago. The unclassified purpose of the project was to develop constant-level balloons for meteorological purposes.

This article originally appeared in the *Skeptical Inquirer* 19 (July/August 1995). Reprinted with permission.

Its classified purpose was to try to develop a way to monitor possible Soviet nuclear detonations with the use of low-frequency acoustic microphones placed at high altitudes. No other means of monitoring the nuclear activities of a closed country like the USSR was yet available, and the project was given a high priority. One of the NYU tasks was the development of constant-level balloons for placing the acoustic microphones aloft. After some preliminary flights in Bethlehem, Pennsylvania, in April 1947, which failed due to high winds, the project moved to New Mexico.

In June and early July 1947, numerous NYU balloon flights were launched from Alamogordo Army Air Field in New Mexico. Some of these flights consisted of very long trains containing up to two dozen neoprene sounding balloons, having a total length of more than 600 feet.

Moore makes a strong case for the hypothesis that NYU Flight #4, which he helped launch on June 4, 1947, was the source of the debris Brazel found on the Foster ranch, and therefore the source of the "Roswell Incident" itself. Many of the materials used in Flight 4 bear striking similarities to pieces of the Roswell debris. A diagram of an earlier, similar flight, Flight #2 (launched April 18, 1947, from Bethlehem, Pennsylvania) appears in figure 1. No such diagram is available for Flight 4; since no altitude data were obtained for it, it was not included in formal NYU reports. However, Moore says the configuration for Flight 4 was quite similar to that shown. The large octahedral objects at top left and bottom middle are radar reflectors, which were used for tracking. Several small aluminum rings for handling the lines are indicated; the "payload" (a sonobuoy) was supported by slightly larger rings. The cluster of neoprene sounding balloons extended for hundreds of feet in flight.

The debris Brazel picked up—and which was later taken to Fort Worth, Texas, for inspection by Brigadier General Roger Ramey, the air force commander there—matches NYU Flight 4 in several different ways. Some of the debris consisted of patches of a smelly, smoky gray, rubberlike material, which is consistent with the neoprene balloons used in NYU Flight 4. Much of the Roswell debris—sticks, metallic paper, and strangely marked tape—is similar to material used for the radar reflectors. When Warrant Officer Irving Newton saw the debris in General Ramey's office, he recognized it as pieces of a radar target. Moore points out that the Ramey photographs show parts of more than one reflector; Flight 4 contained three Signal Corps ML-307B RAWIN targets.

Many witnesses of the debris described tape with flower designs or hieroglyphics on it. Moore recalls that the reinforcing tape used on NYU targets had curious markings. "There were about four of us who were involved in this, and all remember that our targets had sort of a stylized, flowerlike design. I have prepared, in my life, probably more than a hun-

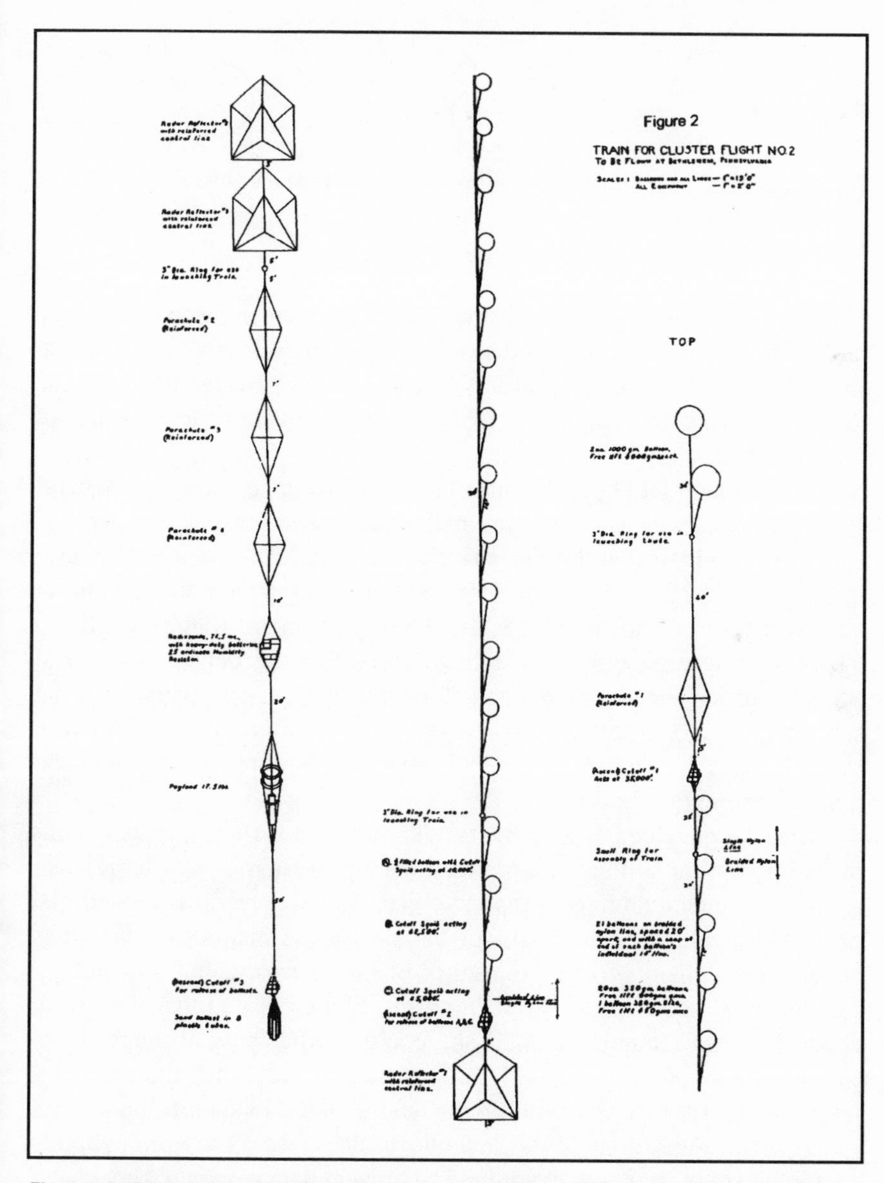

Figure 1. Diagram of balloon train for NYU Flight #2. Charles B. Moore says the configuration for Flight #4 was quite similar.

dred of these targets for flight. And every time I have prepared one of these targets, I have always wondered what the purpose of that tape marking was. But . . . a major named John Peterson, laughed . . . and said 'What do you expect when you get your targets made by a toy factory?' " Figure 2 shows a diagram of Moore's recollection of the flowery design.

Figure 2. "Abstract flowerlike designs" on NYU radar targets.

The radar targets contained small eyelets. Moore showed the NMSR audience a similar target with the eyelets (see photo insert). In an article in the *Roswell Daily Record* on July 9, 1947, rancher Brazel described the debris as having no strings or wire, but as having eyelets for some sort of attachment.

While many UFO proponents claim the wreckage shown in General Ramey's office was just a weather balloon switched for the "real debris," Moore pointed out that the radar targets used by NYU were unlike anything flown in New Mexico before and that "they were not available in Fort Worth to be substituted for the debris in General Ramey's office." Warrant Officer Newton was able to recognize the debris in General Ramey's office because he happened to have used an early version of the same targets while serving as a weatherman in Okinawa. The earlier-model targets Newton used did not have the reinforcing tape with the pinkish-purple flower designs.

Brazel's daughter, Bessie Brazel Schreiber, in a 1979 interview conducted by author William Moore (no relation to Charles B. Moore), described some aluminum ring-shaped objects in the debris that looked like pipe intake collars or the necks of balloons. (The mention of the rings appears in William Moore's transcript of the interview, but was not included in his book *The Roswell Incident.*) She estimated that they were about 4 inches around, and said she could put her hand through them. Charles Moore points out that Flight 4 carried several 3-inch-diameter aluminum rings for assisting with the launching of the balloon train, as well as larger rings used to hold the sonobuoys. These were cut from cylindrical tubing stock, and then chamfered to prevent damage to the ropes.

Sheridan Cavitt, the CIC (Counter-Intelligence Corps) officer who accompanied Major Jesse Marcel to the debris field, described a black box in the wreckage. Moore says the NYU crew routinely packed batteries for the acoustic equipment in black boxes. There has been some speculation that the black box might have been a radiosonde, but Moore pointed out that radiosondes are usually white to prevent absorption of heat.

On June 4, 1947, Flight 4 was launched, and tracked as far as Arabela, New Mexico, only 17 miles from the location of the debris field on the Foster ranch. Flight 4 was still aloft when the batteries ran down, and contact was lost. Brazel reported that he found the debris on the ranch on June 14, 1947, although most UFO proponents put the time of this discovery as a few weeks later, in early July. Brazel didn't take the debris into Roswell until July 7, 1947, by his own account; this date is disputed as well.

Recently, Charles Moore has developed a brand-new line of evidence even further supporting a link between the Roswell Incident and Project Mogul. UFO researcher Kevin Randle recently provided Moore with National Weather Service wind data for early June 1947. Moore, who has lived and breathed atmospheric physics most of his adult life, analyzed this data in detail. His analysis deals with three NYU flights: Flight 4 (June 4, 1947), Flight 5 (June 5), and Flight 6 (June 7). The Weather Service wind data are compatible with what is called a baroclinic weather system moving through the area. As this "trough aloft" slowly passes by, the winds aloft will shift from blowing toward the northeast, then toward the east, and then toward the southeast. At very high altitudes, however, this type of system produces high-level winds in the upper troposphere at cross directions to those at lower levels. Furthermore, the prevailing winds in the stratosphere during the summer months blow toward the west, while those in the transition region just above the tropopause blew toward the northwest during the early part of June 1947. For example, Flight 5 proceeded mainly east as it rose through the troposphere; when it entered the stratosphere, however, it was carried to the northwest. After some balloons burst and Flight 5 descended, it again headed in an easterly direction until it landed.

When Moore used the Weather Service wind data and NYU altitude information to simulate the probable paths of the flights with recorded ground tracks (Flights 5 and 6), his results agreed quite reasonably with the measured balloon paths—Flight 5 drifted mainly to the east, landing near Roswell, while Flight 6 took a more southwesterly route. Moore then extended his analysis to Flight 4, the Roswell candidate. He used the wind data for June 4, 1947, and assumed the flight reached altitudes comparable to those of the subsequent two flights (which were made with very similar balloon trains).

Moore's analysis indicates that after Flight 4 lifted off from Alamogordo, it probably ascended while traveling northeast (toward Arabela), then turned toward the northwest during its passage through the stratosphere, and then descended back to earth in a generally northeast direction. Moore's calculated balloon path is quite consistent with a landing at the Foster ranch, approximately 85 miles northeast of the Alamogordo launch site and 60 miles northwest of Roswell. Furthermore, the debris was strewn along the ground at a southwest-to-northeast angle (as reported by Major Jesse Marcel); this angle is entirely consistent with Moore's analysis.

Charles B. Moore has been repeatedly criticized in the UFO literature for changing some of his earlier statements. He was interviewed for William Moore's book on the Roswell Incident. After hearing Bill Moore's description of the wreckage (including details of supposed 10-inch furrows running some 500 feet), Charlie Moore responded by saying: "Based on the description you gave me, I think that could not have been our balloon." Balloon trains like Flight 4 were far too light to make large furrows in the ground. The issue is not that Charles Moore said the wreckage couldn't have been a balloon—it's that he said his flights couldn't have plowed the alleged "furrows." On another note, Moore and other Mogul participants originally thought the debris Brazel found must have been from one of NYU's polyethylene balloon flights from early July 1947. He held this opinion until just a couple of years ago. These large, transparent polyethylene balloons were used for the first time ever in the summer of 1947 and would have looked strange even to experienced balloon watchers. However, after seeing the reports and photographs from 1947 for the first time, Charles Moore realized that Flight 4 was a much better candidate for the Foster ranch debris than a polyethylene balloon. So he *has* changed his opinions on the incident, but only because better data became available.

Moore's presentation included fascinating details on the background of Project Mogul. He noted that the discovery of the acoustic "duct" between the troposphere and the stratosphere came about as a result of a World War II era analysis of globally propagated sound waves produced by the volcanic explosion of Krakatoa in 1883. In one of their flights, he said the NYU crew attempted (without success) to detect explosions from the British destruction of German installations on the island of Helgoland (off the north German coast) in April 1948. While UFO proponents allege a lack of contemporary references to "Project Mogul Balloon Flights," Moore says the project was so compartmentalized that such references simply may not exist. Any mention of these flights will instead be labeled as NYU constant-level balloon research.

Several UFO authors claim that the wreckage, and possibly alien bodies as well, were secretly flown to Wright Field in Dayton, Ohio for analysis. By coincidence, Moore says he and the rest of the NYU balloon crew stayed over at Wright Field the evening of July 8, 1947, en route back to New Jersey, just as the Roswell story was breaking. Moore says they first learned of the incident while in Dayton, and figured that it was probably caused by one of their recent polyethylene balloon flights.

The September 1994 air force report indicates that the Brazel debris also made its way to Wright Field. During an air force interview of Mogul participant Colonel Albert C. Trakowski, he recalled a July 1947 telephone call from Colonel Marcellus Duffy, who was stationed at Wright Field and was intimately knowledgeable about both Project Mogul and military weather equipment. Duffy told Trakowski that a fellow from New Mexico came to Dayton, woke him up in the middle of the night, and showed him the debris. Colonel Duffy told the fellow, "It looks like some of the stuff you've been launching at Alamogordo."

What is the bottom line on the Roswell Incident, NYU, and Project Mogul? In Moore's words, "When the wind information is coupled with the similarities in the debris described by the eyewitnesses—the balsa sticks, the 'tinfoil,' the tape with pastel, pinkish-purple flowers, the smoky gray balloon rubber with a burnt odor, the eyelets, the tough paper, the four-inch-diameter aluminum pieces and the black box—to the materials used in our balloon flight trains, it appears to me that it would be difficult to exclude NYU Flight 4 as a likely source of the debris that W. W. Brazel found on the Foster ranch in 1947."

UPDATE

The connection of the purported "Roswell Incident" to balloon flights performed by a New York University (NYU) group, in support of the once-secret "Project MOGUL," was first realized by independent UFO researcher Robert G. Todd and later by Karl T. Pflock, and then confirmed by the air force in its detailed analysis published in 1994. Since that time, several UFO proponents have vigorously attacked the NYU/MOGUL explanation. Many have criticized the air force for "admitting" that the 1947 weather balloon explanation was a "lie," and then question MOGUL as just another new lie. However, the original explanation has not been retracted—now, as in 1947, the air force claims that the Roswell Incident was caused by misidentification of *weather equipment.* What is new is that the *source* of this equipment, New York University physics experiments launched from Alamogordo, New Mexico, has finally been identified.

A typical dismissal of the MOGUL hypothesis was put forth in an episode of The Learning Channel's series "Beyond Belief" entitled "Roswell." This television program, aired in 1995, included an interview with C. B. Moore, who discussed his involvement with launches of NYU balloon flights in 1947. Moore was shown saying, "I suspect that our Flight 4, on the 4th of June, was probably the source of the debris the rancher found." The narrator followed this by saying, "Curiously, an examination of the actual flight logs of the time reveals that there isn't any record of the existence of Flight Number 4 on the fourth of June, the one the air force claims explains the Roswell Incident. Moore says no record exists, and no data was entered for unsuccessful flights, yet other flights that were unsuccessful are clearly documented. Discrepancies such as these have led many to doubt the air force's insistence that Project MOGUL is in fact the answer to the Roswell mystery." As these words were read, images of NYU flight logs were displayed, with the words "Flight unsuccessful. Altitude control damaged on launching. 4 lifter balloons, 34 main balloons." Examination of the NYU reports shows that this citation appeared for Flight #6, launched on June 7, 1947.

The resolution of this apparent "discrepancy" hinges on the definition of an "unsuccessful flight." Moore and other NYU personnel did *not* log flights for which no altitude data were obtained. Altitudes were measured either with tracking radar and radar targets (like the ones that puzzled rancher Brazel and Major Marcel), or with the use of radiosondes. If altitude data *were* obtained, then the flight *was* logged, whether or not the acoustic equipment was functioning, because any information on heights actually obtained by the flights was useful. And that's how one type of failure (no altitude data) was not logged, while a different type of failure (no acoustic data from the microphones being used to record sounds of explosions) *could* be logged.

In the same program, New Mexico Congressman Steve Schiff said he was still suspicious of the air force explanation. Schiff stated "I still find it amazing that the United States' top bomber wing, the only wing that was eligible to carry nuclear weapons at that particular time, would not know a weather balloon from a flying saucer, but apparently somebody didn't." Several UFO proponents have echoed this theme, saying that an experienced intelligence officer like Major Marcel simply couldn't have been foolish enough to mistake weather equipment for a spaceship. But Marcel's expertise was in aerial reconnaissance—assessing bomb damage from aerial photographs. He was not familiar with all types of weather equipment, and almost certainly had never previously seen the rarely used radar reflectors. Given the national hysteria about "flying disks" during the summer of 1947, it is not altogether surprising that Marcel thought he might have been presented with the "real thing."

UFO promoters also point to the late Marcel's pronouncements on supposed strange properties of the debris materials as strong evidence for its extraterrestrial origin. But Robert Todd has unearthed several glaring inaccuracies in Marcel's statements to UFO authors (*The KowPflop Quarterly* 1, no. 3, December 8, 1995). For example, Marcel claimed that he was awarded five Air Medals for shooting down five enemy planes. However, official records show that Marcel only received two air medals, and these were for flying on a minimum number of missions, not for shooting down enemy aircraft.

Some of the strongest-sounding MOGUL rebuttals come from Mark Rodeghier and Mark Chesney in their article "What the GAO Found: Nothing About Much Ado," published in the July/August *International UFO Reporter,* regarding the General Accounting Office (GAO) July 1995 report on "Results of a Search for Records Concerning the 1947 Crash Near Roswell, New Mexico." Rodeghier and Chesney claim that "The GAO's lack of success at finding records linking Project MOGUL and Roswell has to undermine the Air Force's purported explanation. . . . In our view, the GAO report makes it less likely that Project MOGUL was the source of the Roswell debris, since no records were located to support that conjecture." While the GAO report does not explicitly mention MOGUL, it does refer to the air force Roswell report, both in Table 1 and in Appendix VII. And the air force report clearly *does* mention the connection of MOGUL to the unclassified NYU balloon flights.

C. B. Moore has noted that "MOGUL" was a code name used by certain groups within the Army Air Force, and that he never even heard the term until the 1990s. In Moore's words, "Since the NYU flights were never identified, prior to 1994, as being made in support of Project MOGUL, there are no records in the archives associating them with that project (and, of course, there are no records on 'Project MOGUL balloon flights' because there were never any flights with that name). On the other hand, a rather complete file on the New York University Balloon Project under Contracts W28-099-ac-241, AF 19(122)-145 and AF 19(122)-633 is stored at Hanscom Air Force Base in Bedford MA, together with the NYU progress and technical reports. As in all searches, blind alleys are encountered until the searcher knows where to look" (letter to Philip J. Klass, September 25, 1995).

Several NYU technical reports are included in their entirety in the publicly available version of the air force report, entitled "The Roswell Report, Fact Versus Fiction in the New Mexico Desert," and published in 1995. This large volume may be obtained from the U.S. Government Printing Office, Superintendent of Documents, Mail Stop: SSOP, Washington, D.C., 20402-9328. This work also includes a paper entitled "Controlled-Altitude Free Balloons" by A. F. Spilhaus, C. F. Schneider, and C. B. Moore, originally published in the *Journal of Meteorology* 5 (1948): 130–37. One of the figures from this paper shows the flight path of a balloon assembly (NYU Flight #11) that crossed over the Sacramento mountains en route from Alamogordo to a spot near Roswell on July 7, 1947, the day the Roswell Incident story broke. The *Journal of Meteorology* paper, which can be found in hundreds of libraries across the country, does not mention "MOGUL," but it certainly *does* provide definitive proof that NYU balloons were flying over Roswell in the summer of 1947.

I don't think the UFO proponents should be taken to task for having doubts about the authenticity of the NYU connection to the Roswell Incident. In fact, I commend them for being skeptical and cautious, and urge them to apply these same stringent standards to reports that the Roswell Incident involved an alien spaceship. Fifty years after the "Incident," we only have conflicting stories on the location of the crash "site" (there are now up to six spots competing for the honor), on the number of dead aliens observed (anywhere from two to five), and numerous other discrepancies. What we *don't* have is one piece of solid, physical evidence that Roswell involves something not of this Earth. At present, the hypothesis that Roswell was simply a brief case of mistaken identity, in which some weather equipment was erroneously identified as a "flying disk," is strongly supported by the weight of evidence.

13. THE GAO ROSWELL REPORT AND CONGRESSMAN SCHIFF

Philip J. Klass

An eighteen-month search completed in 1995 of United States government documents—including once highly classified minutes of meetings of the National Security Council—conducted by the General Accounting Office at the request of Representative Steven Schiff of New Mexico, failed to find anything to indicate that the government recovered a crashed flying saucer north of Roswell, New Mexico, in mid-1947. The investigation was conducted by GAO's National Security and International Affairs Division, which has access to the most highly classified information.

The GAO investigators discovered *nothing* to challenge the conclusions of a 1994 report by the United States Air Force, based on its own extensive investigation. The air force concluded that the unusual material recovered from a ranch north of Roswell was debris from a train of balloons, radar tracking targets, and other devices associated with a then top-secret Project Mogul. (*Skeptical Inquirer* 19, no. 1, January–February 1995 [chapters 10 and 11 in this book].)

But you could get a vastly different impression from the news release issued by Schiff on July 28, 1995, which formed the basis of many news media stories, including one filed by the Associated Press. Schiff's two-page news release carried the headline: "SCHIFF RECEIVES, RELEASES ROSWELL REPORT (missing documents leave unanswered questions)."

The release began: "Congressman Steve Schiff today released the

---

This article originally appeared in the *Skeptical Inquirer* 19 (November/December 1995). Reprinted with permission.

General Accounting Office (GAO) report detailing the results of a records audit relating to events surrounding a crash in 1947, near Roswell, N.M., and the military response. The twenty-page report is the result of constituent information requests to Congressman Schiff and the difficulty he had getting answers from the Department of Defense in the now forty-eight-year-old controversy.

"Schiff said important documents, which may have shed more light on what happened at Roswell, are missing. 'The GAO report states that the *outgoing* messages from Roswell Army Air Field (RAAF) *for this period of time* were destroyed without proper authority' [emphasis added]. Schiff pointed out that these messages would have shown how military officials in Roswell were explaining to their superiors exactly what happened."

Based on the wording of Schiff's news release, one might conclude that the "missing" outgoing RAAF teletype messages were only for a brief period in early July of 1947. But the GAO reports its auditors were unable to locate any outgoing RAAF messages for a *three-year period* extending from October 1946 through December 1949.

During an interview with Schiff in his Washington office on July 29, 1995, I noted that Pentagon officials first learned from news wire service reports—rather than official channels—that the Roswell Army Air Field (RAAF) had announced recovery of one of the then mysterious "flying disks." Because the flying disks might have been Soviet spy vehicles, I asked the Congressman if it would not have been more logical for Pentagon officials to have called the RAAF base commander on the telephone rather than take time to compose and transmit a teletype inquiry. Schiff replied: "I think they would have done it by both."

Schiff's news release failed to mention that when the GAO examined once highly classified minutes of meetings of the National Security Council for 1947 and 1948, it found no mention of the Roswell incident. I asked Schiff: "If the U.S. government had recovered an alien spacecraft in New Mexico in July of 1947, do you not believe that that extraordinary event would have been discussed at National Security Council meetings?"

Schiff responded: "I would have to say, but let me say first, my endeavor has *never* been to look for UFOs or aliens as such. My endeavor has been to look to see what was in the government records insofar as they could be reconstructed at this point, which after 50 years is problematic. And I went to the GAO because the Department of Defense would not be cooperative in that regard—in fact, I believe, gave me the run-around when I requested the information."

When I pressed Schiff to answer my question, he responded: "It would be such an unusual event . . . that I'm not sure how it would be handled and even if it were presented to the national leaders and National Security Council, I'm not sure I would necessarily say that you could say how they would handle the minutes of such a meeting." In other words, Schiff is uncertain whether recovery of an alien spacecraft—which could be the precursor of an attack on Earth—would be reported to and discussed by the president and National Security Council. And even if discussed, Schiff is unsure whether there would be any mention of the incident in any of the highly classified minutes of NSC meetings.

The GAO report included a copy of an outgoing teletype message from the Dallas bureau of the Federal Bureau of Investigation to FBI headquarters, sent at 6:17 P.M. on July 8, 1947, that read:

> EIGHTH AIR FORCE, TELEPHONICALLY ADVISED THIS OFFICE THAT AN OBJECT PURPORTING TO BE A FLYING DISC WAS RECOVERED NEAR ROSWELL, NEW MEXICO, THIS DATE. THE DISC IS HEXAGONAL IN SHAPE AND WAS SUSPENDED FROM A BALLOON BY CABLE. . . . FURTHER ADVISED THAT THE OBJECT RESEMBLES A HIGH ALTITUDE WEATHER BALLOON WITH A RADAR REFLECTOR. . . .

The GAO report also includes a copy of the "Combined History, 509th Bomb Group and Roswell Army Air Field, 1 July 1947 to 31 July 1947." This once-classified document reports: "The Office of Public Information was kept quite busy during the month answering inquiries on the 'flying disc' which was reported to be in the possession of the 509th Bomb Group. The object turned out to be a radar tracking balloon."

The congressman's news release briefly summarizes these documents but dismisses their importance in the following words: "Even though the weather balloon story has since been discredited by the U.S. Air Force, Schiff suggested that the authors of those communications may have been repeating what they were told rather than consciously adding to what some believe is a 'cover-up.'"

The congressman was quoted as saying: "At least this effort caused the Air Force to acknowledge that the crashed vehicle was no weather balloon. That explanation never fit the fact of high military security used at the time." Clearly, Schiff has not carefully studied the 1994 air force report and seemingly believes that RAAF's action in issuing a news release saying it had recovered a flying disk can be characterized as "high military security."

The original air force identification of the debris—discovered by rancher "Mac" Brazel—as the remnants of a weather balloon and radar tracking target, was made on July 8, 1947, in the office of Brigadier General Roger Ramey, Eighth Air Force commander at Fort Worth by Weather Officer Irving Newton. At the time, neither officer had the security clearance necessary to know about a then top secret experimental program, called Project Mogul, which was then under way at the Alamogordo Army Air Field in New Mexico. The project's objective was to explore the feasibility of using high-altitude balloons outfitted with acoustic sensors to detect when the Soviets tested their first nuclear weapon.

On June 4, 1947, a cluster ("train") of more than twenty weather balloons with multiple radar targets was launched from the Alamogordo Army Air Field and was tracked to within seventeen miles of the Brazel ranch before radar contact was lost. Brazel discovered the unusual debris 10 days later. (See *Skeptical Inquirer* 19, no. 4, July–August 1995, p. 15 [This article is chapter 12 of this volume.].) The description of the debris given by rancher Brazel on July 8, 1947, in the offices of the *Roswell Daily Record,* and recent recollections of his daughter Bessie, who helped her father collect the debris, indicate that the debris came from this launch of a train of ordinary weather balloons and associated equipment.

The recent investigation by the air force into claims of a crashed flying saucer near Roswell was initiated in early 1994 in response to a GAO request, and its report was released in September 1994. The air force investigation officially uncovered the link to Project Mogul, which had been discovered about two years earlier by UFO researcher Robert Todd, and more recently by UFO researcher Karl Pflock. (Pflock's wife, Mary Martinek, is Schiff's chief of staff and his liaison with the GAO for its Roswell investigation.) The 1994 air force report, page 21, states that "the most likely source of the wreckage recovered from the Brazel Ranch was from one of the Project Mogul balloon trains."

In December 1992, shortly after Pflock had launched his own personal investigation into the Roswell incident, he supplied Schiff with a 130-page briefing paper on the subject. Three months later, Schiff wrote to then Defense Secretary Les Aspin seeking a "definitive explanation of what transpired and why." Schiff's letter said that based on (alleged) witness testimony, "the balloon explanation was a cover story" and that "federal authorities sought to intimidate witnesses and their families into silence," according to an article in the January 14, 1994, *Albuquerque Journal.*

Since then, Pflock's several-year investigation has convinced him that "at least the great majority if not all" of the debris found by Brazel was wreckage from the cluster of balloons, radar targets and instruments launched from Alamogordo on June 4, 1947. In Pflock's invited talk to New Mexicans for Science and Reason in August, he said he thinks that "most reasonable people will agree" and that he believes that the evidence is "fairly conclusive."

When the *Albuquerque Journal* published an article by its Washington correspondent, Richard Parker, who interviewed Schiff about the GAO report, the article carried the headline: "Schiff: Roswell UFO a Balloon." This prompted Schiff to challenge the accuracy of Parker's article in a letter published in the newspaper August 14, 1995.

Wrote Schiff: "With the sole exception of rejecting the original military explanation of a crashed 'weather balloon,' which the Air Force now disavows, I have never stated any conclusion about the Roswell crash. . . . Of course, the 1994 Air Force explanation is a possible answer. . . ." Schiff said that the GAO inquiry, which he generated, "has had some notable results in addition to forcing the Air Force to change its story."

Schiff's letter also said:

• "Two documents were uncovered which refer to a 'radar tracking device,' (which means weather balloon) though the writers at the time could merely have been repeating what they were told."

• "Agencies, including the CIA, stated for the first time that they do not have information on the Roswell incident."

• "Perhaps most significantly, documents most likely to contain helpful information, the military's outgoing messages, were not found. It was estimated they were destroyed over 40 years ago without proper authority. This means the military cannot explain who destroyed the records, or why."

Schiff's published letter concluded: "Yet, from this, Parker manages to conclude for me that the Air Force came clean. His inference is clearly out of this world."

In early 1994, when it was first disclosed that Schiff had asked the GAO to investigate the Roswell incident and he was interviewed by the *Albuquerque Journal,* the newspaper reported that Schiff said "he doesn't believe a UFO was recovered at the ranch." The article quoted Schiff as saying: "If I had to guess, I would say some kind of military experiment."

Because Schiff's guess proved to be remarkably prescient, I asked him if the GAO report and the 1994 air force report had increased his earlier-stated belief that the debris discovered by rancher Brazel was *not* from a UFO. He responded: "I think you're centering too much on my beliefs in the matter," but he acknowledged that the Project Mogul explanation "could well be what actually happened."

Schiff predicted that "the GAO report will not change anybody's mind" about whether the government recovered a crashed flying saucer in 1947. "People can make their own conclusions and that was my goal all along and I have accomplished that goal," Schiff said.

Schiff has had extensive media exposure as a result of his Roswell activities, including appearances on numerous local and network television shows. He has twice appeared on Larry King's show (CNN) to discuss his Roswell efforts.

Schiff said he is convinced "that people have a right to information from their government on any subject—with the notable exception of [information affecting national] security." This prompted me to ask if Schiff planned to seek congressional hearings on the all-important but still unresolved issue of whether the United States government is involved in a UFO coverup. Schiff said, "I have no intention of taking it further."

14. THE "ROSWELL FRAGMENT"—CASE CLOSED

David E. Thomas

It was said to have come from outer space. On Sunday, March 24, 1996, a small piece of strange, swirly-patterned metal was delivered to the Roswell International UFO Museum and Research Center in Roswell, New Mexico. The man who turned in the fragment of metal to the museum, Blake Larsen, had been told it was retrieved from the 1947 crash of an alien craft near Roswell. But on Friday, September 6, it was revealed that the fragment wasn't made on another planet—it came from St. George, Utah. Here is the story of the fragment that has gained international attention for the last six months.

Shortly after receiving the piece of metal, the museum announced plans to have it scientifically tested. Museum official Max Littel said, "If some metallurgist says there is nothing in the book like this, and he has got all his degrees and is an expert source, then we are home free. This is it."

When I heard of the fragment on local television, I promptly contacted C. B. Moore, professor emeritus of physics at New Mexico Tech. Back in 1947, Moore launched the balloon train now widely thought to have precipitated the actual "Roswell Incident" (*Skeptical Inquirer,* July/August 1995; chapter 12 in this book). Moore then spoke with museum board member Miller Johnson and was invited to join in the first inspection of the fragment, scheduled for the following day (March 29), at the Bureau of Mines at New Mexico Tech in Socorro. Department manager Chris McKee carried out the analysis, while Roswell police chief Ray Mounts recorded the procedure. The results of the X-ray fluores-

---

This article originally appeared in the *Skeptical Inquirer* 20 (November/December 1996). Reprinted with permission.

cence measurements indicated the metal was about 50 percent Cu (copper) and 50 percent Ag (silver) on the front side, and 87 percent Ag and 12 percent Cu on the back side, with 1 percent other trace elements.

At about that time, I came across a statement by astronomer Carl Sagan in the March/April 1996 *Skeptical Inquirer* regarding testing of purported alien artifacts such as implants. Sagan noted that none had been observed to have unusual isotopic content. I did a little research and found that most isotopes of copper decay quickly, but two are stable: Cu-63 and Cu-65. No matter where copper is found on Earth, it always consists of the same percentages of these isotopes. But heavy elements like copper are forged by a variety of thermonuclear events in red giants or supernovae, and thus the ratios of various isotopes will most likely vary from star to star. I passed the suggestion of isotope testing to Moore, who passed it on to Johnson, who began setting up the tests in May.

The isotopic analyses took place at Los Alamos National Laboratory (LANL) on August 1 and 2, 1996. The museum paid $725 for the work. E. Larry Callis of the Chemical Metallurgical Research division performed the tests, which were taped and photographed by LANL personnel and Johnson. Pieces of both the original and a second "fragment" were placed in a mass spectrometer and measured. The Cu-63 results for fragments 1 and 2 were 69.127 percent and 69.120 percent, respectively. A piece of normal, refined copper, tested as a control, had a value of 69.129 percent Cu-63. The accepted value for Cu-63 is 69.174 percent. And so, the copper was not found to deviate significantly from Earthly isotopic ratios. A similar result was obtained for the silver, which contains 51.840 percent Ag-107 and 48.160 percent Ag-109.

The results of the LANL test were mentioned in *NMSR Reports* (the New Mexicans for Science and Reason newsletter), where they caught the eye of *Albuquerque Journal* reporter John Fleck. Fleck's August 13 report on the isotopic analyses caught the eye of someone in St. George, Utah, who called Fleck to tell him where the fragment really came from.

It turns out the fragment was a piece of leftover material created by artist Randy Fullbright. Fullbright uses an ancient Japanese technique (translated as "wood-grain-metal" by Johnson's wife, Marilyn) to create the swirly patterns of copper and silver. Fullbright gave the material to a gallery owner, yet to be named, and did not identify it as anything other than scraps from his own artwork. The gallery owner gave it to Blake Larsen, who was leaving St. George to move to Roswell. The gallery owner told Larsen that the fragment was "found near Roswell in 1947." When Larsen got to Roswell, he gave the piece to the museum. Fleck confirmed the details of the story with Fullbright and with Larsen, as well as with museum officials. He published the story in an article, "Artist: Fragment Is Bogus," appearing on the front page of the September 6 *Albuquerque Journal.*

And so ends the saga of the Roswell fragment. The museum officials took the risk of having the debris definitively tested, and scientists reciprocated with serious, careful measurements. The specimen turned out to be Earthly this time. If extraterrestrial landings are ever to be confirmed in the future, it may be with experiments like these.

Part Three
ROSWELL AND THE "ALIEN AUTOPSY"

15.
"ALIEN AUTOPSY" HOAX
Joe Nickell

It keeps going and going and. . . . The Roswell crashed-saucer myth has been given renewed impetus by a controversial television program "Alien Autopsy: Fact or Fiction?" that purports to depict the autopsy of a flying saucer occupant. The "documentary," promoted by a British marketing agency that formerly handled Walt Disney products, was aired August 28 and September 4, 1995, on the Fox television network. Skeptics, as well as many UFOlogists, quickly branded the film used in the program a hoax.

"The Roswell Incident," as it is known, is described in several controversial books, including one of that title by Charles Berlitz and William L. Moore. Reportedly, in early July 1947, a flying saucer crashed on the ranch property of William Brazel near Roswell, New Mexico, and was subsequently retrieved by the United States government (Berlitz and Moore 1980). Over the years, numerous rumors, urban legends, and outright hoaxes have claimed that saucer wreckage and the remains of its humanoid occupants were stored at a secret facility—e.g., a (nonexistent) "Hangar 18" at Wright Patterson Air Force Base—and that the small corpses were autopsied at that or another site (Berlitz and Moore 1980; Stringfield 1977). [See chapter 13, this volume.]

UFO hoaxes, both directly and indirectly related to Roswell, have since proliferated. For example, a 1949 science fiction movie, *The Flying Saucer,* produced by Mikel Conrad, purported to contain scenes of a captured spacecraft; an actor hired by Conrad actually posed as an FBI agent

---

This article originally appeared in the *Skeptical Inquirer* 19 (November/December 1995). Reprinted with permission.

and swore the claim was true. In 1950, writer Frank Scully reported in his book *Behind the Flying Saucers* that the United States government had in its possession no fewer than three Venusian spaceships, together with the bodies of their humanoid occupants. Scully, who was also a *Variety* magazine columnist, was fed the story by two confidence men who had hoped to sell a petroleum-locating device allegedly based on alien technology. Other crash-retrieval stories followed, as did various photographs of space aliens living and dead: One gruesome photo portrayed the pilot of a small plane, his aviator's glasses still visible in the picture (Clark 1993).

Among recent Roswell hoaxes was the MJ-12 fiasco, in which supposed top secret government documents—including an alleged briefing paper for President Eisenhower and an executive order from President Truman—corroborated the Roswell crash. Unfortunately, document experts readily exposed the papers as inept forgeries (Nickell and Fischer 1990).

Sooner or later, a Roswell "alien autopsy" film was bound to turn up. That predictability, together with a lack of established historical record for the bizarre film, is indicative of a hoax. So is the anonymity of the cameraman. But the strongest argument against authenticity stems from what really crashed at Roswell in 1947. According to recently released air force files, the wreckage actually came from a balloon-borne array of radar reflectors and monitoring equipment launched as part of the secret Project Mogul and intended to monitor acoustic emissions from anticipated Soviet nuclear tests. In fact, materials from the device match contemporary descriptions of the debris (foiled paper, sticks, and tape) given by rancher Brazel's children and others (Berlitz and Moore 1980; Thomas 1995).

Interestingly, the film failed to agree with earlier purported eyewitness testimony about the alleged autopsy. For example, multiple medical informants described the Roswell creatures as lacking ears and having only four fingers with no thumb (Berlitz and Moore 1980), whereas the autopsy film depicts a creature with small ears and five fingers in addition to a thumb. Ergo, either the previous informants are hoaxers, or the film is a hoax, or both.

Although the film was supposedly authenticated by Kodak, only the leader tape and a single frame were submitted for examination, not the entire footage. In fact, a Kodak spokesman told the *Sunday Times* of London: "There is no way I could authenticate this. I saw an image on the print. Sure it could be old film, but it doesn't mean it is what the aliens were filmed on."

Various objections to the film's authenticity came from journalists, UFO researchers, and scientists who viewed the film. They noted that it bore a bogus, nonmilitary codemark ("Restricted access, AO1 classification") that disappeared after it was criticized; that the anonymous photographer's alleged military status had not been verified; and that the injuries sustained by the extraterrestrial were inconsistent with an air crash. On the basis of such objections, an article in the *Sunday Times* of London advised: "RELAX. The little green men have not landed. A much-hyped film purporting to prove that aliens had arrived on earth is a hoax" (Chittenden 1995).

Similar opinions on the film came even from prominent Roswell-crash partisans: Kent Jeffrey, an associate of the Center for UFO Studies and author of the "Roswell Declaration" (a call for an executive order to declassify any United States government information on UFOs and alien intelligence) stated "up front and unequivocally there is no (zero!!!) doubt in my mind that this film is a fraud" (1995). Even arch Roswell promoter Stanton T. Friedman said: "I saw nothing to indicate the footage came from the Roswell incident, or any other UFO incident for that matter" ("Alien or Fake?" 1995).

Still other critics found many inconsistencies and suspicious elements in the alleged autopsy. For example, in one scene the "doctors" wore white, hooded anticontamination suits that could have been neither for protection from radiation (elsewhere the personnel are examining an alien body without such suits), nor for protection from the odor of decay or from unknown bacteria or viruses (either would have required some type of breathing apparatus). Thus it appears that the outfits served no purpose except to conceal the "doctors'" identities.

American pathologists offered still more negative observations. Cyril Wecht, former president of the National Association of Forensic Pathologists, seemed credulous but described the viscera in terms that might apply to supermarket meat scraps and sponges: "I cannot relate these structures to abdominal contexts." Again, he said about contents of the cranial area being removed: "This is a structure that must be the brain, if it is a human being. It looks like no brain that I have ever seen, whether it is a brain filled with a tumor, a brain that has been radiated, a brain that has been traumatized and is hemorragic. . . ." (Wecht 1995). Much more critical was the assessment of nationally known pathologist Dominick Demaio who described the autopsy on television's "American Journal" (1995): "I would say it's a lot of bull."

Houston pathologist Ed Uthman (1995) was also bothered by the unrealistic viscera, stating: "The most implausible thing of all is that the 'alien' just had amorphous lumps of tissue in 'her' body cavities. I cannot fathom that an alien who had external organs so much like ours could not have some sort of definitive structural organs internally." As well, "the prosectors did not make an attempt to arrange the organs for demonstration for the camera." Uthman also observed that there was no body block, a basic piece of equipment used to prop up the trunk for examination and the head for brain removal. He also pointed out that "the prosector used scissors like a tailor, not like a pathologist or surgeon" (pathologists and surgeons place the middle or ring finger in the bottom scissors hole and use the forefinger to steady the scissors near the blades). Uthman further noted that "the initial cuts in the skin were made a little too Hollywood-like, too gingerly, like operating on a living patient" whereas autopsy incisions are made faster and deeper. Uthman faulted the film for lacking what he aptly termed "technical verisimilitude."

The degree of realism in the film has been debated, even by those who believe the film is a hoax. Some, like Kent Jeffrey (1995), thought the autopsy was done on a specially altered human corpse. On the other hand, many—including movie special effects experts—believed a dummy had been used. One suspicious point in that regard was that significant close-up views of the creature's internal organs were consistently out of focus ("Alien or Fake?" 1995).

"American Journal" (1995) also featured a special effects expert who doubted the film's authenticity and demonstrated how the autopsy "incisions"—which left a line of "blood" as the scalpel was drawn across the alien's skin—could easily have been faked. (The secret went unexplained but probably consisted of a tube fastened to the far side of the blade.)

In contrast to the somewhat credulous response of a Hollywood special effects filmmaker on the Fox program, British expert Cliff Wallace of Creature Effects provided the following assessment:

> None of us were of the opinion that we were watching a real alien autopsy, or an autopsy on a mutated human which has also been suggested. We all agreed that what we were seeing was a very good fake body, a large proportion of which had been based on a lifecast. Although the nature of the film obscured many of the things we had hoped to see, we felt that the general posture and weighting of the corpse was incorrect for a body in a prone position and had more in common with a cast that had been taken in an upright position.
>
> We did notice evidence of a possible molding seam line down an arm in one segment of the film but were generally surprised that there was little other evidence of seaming which suggests a high degree of workmanship.
>
> We felt that the filming was done in such a way as to obscure details rather than highlight them and that many of the parts of the autopsy that would have been difficult to fake, for example the folding back of the chest flaps, were avoided, as was anything but the most cursory of limb movement. We were also pretty unconvinced by the lone removal sequence. In our opinion the insides of the creature did not bear

> much relation to the exterior where muscle and bone shapes can be easily discerned. We all agreed that the filming of the sequence would require either the use of two separate bodies, one with chest open, one with chest closed, or significant redressing of one mortal. Either way the processes involved are fairly complicated and require a high level of specialized knowledge.

Another expert, Trey Stokes—a Hollywood special effects "motion designer" whose film credits include *The Abyss, The Blob, Robocop Two, Batman Returns, Gremlins II, Tales from the Crypt,* and many others—provided an independent analysis at CSICOP's request. Interestingly, Stokes's critique also indicated that the alien figure was a dummy cast in an upright position. He further noted that it seemed lightweight and "rubbery," that it therefore moved unnaturally when handled, especially in one shot in which "the shoulder and upper arm actually are floating rigidly above the table surface, rather than sagging back against it" as would be expected (Stokes 1995).

CSICOP staffers (Executive Director Barry Karr, *Skeptical Inquirer* Assistant Editor Tom Genoni, Jr., and I) monitored developments in the case. Before the film aired, CSICOP issued a press release, briefly summarizing the evidence against authenticity and quoting CSICOP Chairman Paul Kurtz as stating: "The Roswell myth should be permitted to die a deserved death. Whether or not we are alone in the universe will have to be decided on the basis of better evidence than that provided by the latest bit of Roswell fakery. Television executives have a responsibility not to confuse programs designed for entertainment with news documentaries."

REFERENCES

"Alien or Fake?" 1995. *Sheffield Star* (England), August 18.

"American Journal." 1995. September 6.

Berlitz, Charles, and William L. Moore. 1980. *The Roswell Incident.* New York: Grosset and Dunlap.

Chittenden, Maurice. 1995. "Film that 'Proves' Aliens Visited Earth Is a Hoax." *Sunday Times* (London), July 30.

Clark, Jerome. 1993. "UFO Hoaxes." In *Encyclopedia of Hoaxes,* edited by Gordon Stein, 267–78. Detroit: Gale Research.

Jeffrey, Kent. 1995. "Bulletin 2: The Purported 1947 Roswell Film." Internet, May 26.

Kurtz, Paul. 1995. Quoted in CSICOP press release, "Alien Autopsy: Fact or Fiction? Film a Hoax Concludes Scientific Organization." April 25.

Nickell, Joe, and John F. Fischer. 1990. "The Crashed-Saucer Forgeries." *International UFO Reporter* (March/April 1990): 4–12.

Stokes, Trey. 1995. Personal communication, August 29–31.

Stringfield, Leonard H. 1977. *Situation Red: The UFO Siege.* Garden City, N.Y.: Doubleday. (See pages 84, 177–79.)

Thomas, Dave. 1995. "The Roswell Incident and Project Mogul." *Skeptical Inquirer* 19, no. 4 (July–August): 15–18.

Uthman, Ed. 1995. "Fox's 'Alien Autopsy': A Pathologist's View." Usenet, sci.med.pathology, September 15.

Wallace, Cliff. 1995. Letter to Union Pictures, August 3, quoted in Wallace's letter to Graham Birdsall, *UFO Magazine,* August 16, quoted on ParaNet, August 22.

Wecht, Cyril. 1995. Quoted on "Alien Autopsy: Fact or Fiction?" Fox Network, August 28 and September 4.

16.
"ALIEN AUTOPSY" SHOW-AND-TELL
C. Eugene Emery, Jr.

There's nothing more maddening than having someone invite you to make up your own mind about a controversy, only to have them refuse to give you the tools to do it.

That's precisely what the Fox television network did August 28 and September 4, 1995, when it presented a one-hour special "Alien Autopsy: Fact or Fiction?" that was billed as the network premiere of a seventeen-minute film purporting to be the autopsy of a space creature found near Roswell, New Mexico, in 1947. [See also chapter 13 in this volume.]

Instead of simply showing the seventeen minutes, viewers got to see maybe three, four, or five minutes of footage chopped up into MTV-sized snippets that were repeated throughout the hour.

Instead of a tough skeptical analysis of a film that has been kept tightly under wraps by its owner, executive producer Robert Kiviat—whose résumé includes being a coordinating producer on Fox's pseudoscience newsmagazine program "Encounters"—"Alien Autopsy" tended to showcase interviews from people who seemed convinced that the footage was either real, or a complicated hoax that would have been extremely difficult to pull off.

"Alien Autopsy" was far from one-sided. Kiviat repeatedly had the host, "Star Trek" actor Jonathan Frakes, note that the movie could be a hoax, and Kiviat addressed some key criticisms. But other important criticisms were muted, ignored, taken out of context, or simply brushed aside.

This article originally appeared in the *Skeptical Inquirer* 19 (November/December 1995). Reprinted with permission.

It's understandable that some people would be impressed by the film. The snippets the producers chose to air looked convincing in many ways. Scalpels seemed to cut flesh. A skin flap from the skull seemed to be pulled over the face. Dark innards were removed from the brain area and the body cavity, and placed into pans. The tools and equipment seemed to be from the right era.

Yet when it comes to exposing a clever fraud, the devil is in the details.

By failing to show the entire film, one was left to wonder whether Fox was leaving out the portions that might have flagged the movie as bogus.

"Alien Autopsy" comes at a difficult time for UFO enthusiasts. Today's cutting-edge UFO tales have become so extraordinary, they're often met with derision, even by people in the increasingly sensationalist media.

That's why the focus seems to have shifted to Roswell, where the details are still intriguing enough to fire the imagination, and the facts and recollections have been polished bright by the passage of time. With its simple tale of a crashed saucer, a few space aliens, and a government cover-up, the Roswell story seems far more plausible (relatively speaking) than today's tales of aliens passing through walls, millions of Americans being abducted by sex-obsessed space creatures, and extraterrestrials who create alien-human babies.

UFO believers thought they had the Roswell affair pretty well figured out. "Alien Autopsy" has shaken things up because the images in the film don't always conform to the picture the believers have painstakingly constructed over the years. The creature on the autopsy table is tall, its eyes are too small, it has too many fingers and toes, and it looks too humanlike, complete with humanlike ears and toenails.

Some enthusiasts had expressed the fear that "Alien Autopsy" would discredit some of the work that has gone into uncovering the truth at Roswell. Such fears may be justified. In the media, it's the images, not facts, that shape public attitudes and debates these days. Long after people have forgotten the details of a Roswell book or article, they're going to remember the video of this six-fingered "alien" undergoing an "autopsy."

The film snippets that were shown raised all kinds of questions, and provided few answers. Some examples:

• One small part of the film shows someone making a cut in the skin along the neck. Did the full-length film include the showing of any dissection of the cut area? Was this cutting of skin simply done for effect, possibly with a trick knife that makes a glistening mark on the body that appears to be the blood from an incision?

• One section of the film shows an intact body (except for a large leg wound). Another shows the thorax and abdomen cut open. Were there any steps in between, or did possible hoaxers making the film simply cut open a latex dummy, dump animal guts inside, and pretend to take them out?

• There were film clips of organs, such as the brain, being removed. But organs can't be pulled from a body like pieces in a jigsaw puzzle. They're held in position by sometimes-tough connective tissue that must first be cut away. The film snippets on "Alien Autopsy" showed no evidence of that type of dissection. That flaw—if it is a flaw—was most obvious when the doctor plucked the dark covering off the eye. Unless these were simply extraterrestrial contact lenses, a piece of the eye isn't going to come away that easily without some connective tissue being sliced first.

• Where was everybody? How many people would turn down the chance to watch the historic autopsy of a creature from another world? Yet there were only two people in this room, in addition to the cameraman.

• Why did the person watching from behind the glass partition, and not in the room, need to be suited up?

• For such an extraordinary autopsy, why did there seem to be so little effort to document it? There was no attempt to weigh or label the specimens, and there were just a few shots of someone putting data on a single sheet of paper.

• Why was the supposedly experienced cameraman—who also claims to have been present when three alien creatures were found—trying to take close-ups that invariably made the film go out of focus? Good photographers know when they're getting too close to their subject and need to switch to a lens with a more appropriate focal length.

The fact is, an autopsy on a creature this extraordinary wouldn't be done the way this one was. The being would have been turned over so the back could be examined (in fact, the "doctors" seemed reluctant to move the body much at all). The skin would have been carefully stripped away to examine the pattern of the musculature. The origin and insertion of individual muscles would have been documented. Samples would have been taken, weighed, recorded, and photographed. Only then would the people behind the protective hoods have gone deeper into the gut, repeating the documentation process.

When critics have questioned the quick removal of the black sheath on the eyes, the argument has been made that this was the third or fourth alien autopsied, so the procedure was becoming easier. The argument doesn't wash. Unless this was one of scores of alien bodies, researchers would want to handle each case with excruciating care so they could compare and contrast the individuals.

Unfortunately, the people who were skeptical of the film—ironically, including people prominent in the UFO movement—were given little time and almost no opportunity to explain their skepticism, making them appear to be little more than debunkers. Kent Jeffrey, who argued months earlier that the film is a hoax, only got to predict that it will probably eventually be exposed as a fraud. The criticisms of one Hollywood filmmaker, who thought the movie was bogus, were quickly countered by a cameraman from the era who said it wasn't surprising that this autopsy cameraman would allow his view to be blocked or parts of the movie to be out of focus.

Then there were things the show didn't tell viewers.

"Alien Autopsy" quoted Laurence Cate of Kodak, who said the markings on the film indicate it was manufactured in 1927, 1947, or 1967. The program didn't make it clear that Cate is not an expert in authentication, according to the *Sunday Times* of London.

Paolo Cherchi Usai, senior curator at George Eastman House, a photography museum, based his observation that the film would be difficult to fabricate on seeing the seventeen minutes of film and about five frames of leader film that carried no date coding and was supposedly clipped from the beginning of one of the rolls of film. Conclusive tests on the film had yet to be done.

The Hollywood special effects team led by Stan Winston gave the most impressive testimonial. But I got the impression they were being asked to gauge the difficulty of staging a bogus alien autopsy back in 1947. Winston and his associates said the special effects were good, even by today's standards, but from the clips shown on "Alien Autopsy," this television program didn't seem to come close to rivaling the quality of films you could rent in any video store.

The bottom line is that if the film is legitimate and this is the first solid evidence of life on other planets, it deserves real authentication, not the casual checking the program provided.

Independent experts need to pinpoint the date of the frames, then examine all the reels to be sure the entire film has the same date code. For all we know, most of the film is from contemporary stock. Checking the whole film would dramatically narrow the range of possibilities for a hoax.

The cameraman needs to be identified and questioned to confirm that he exists, that he was in the military, and that he really was the cameraman. There's been talk that he wants to avoid being prosecuted by the government for keeping a copy of the film all these years. That's claptrap. If the film is a hoax, why would the government bother him? If the film is real, dragging a more-than-eighty-year-old military veteran into court would be an admission by the government that the footage is real, and that would spark some tough questions about who or what was on that examining table. The government, not the photographer, would be on the hot seat.

But instead of insisting on authentication first, Fox seemed intent on milking the movie for every penny possible. The network repeated the program one week after its original showing and tried to drum up renewed interest for the rerun by promising more footage from the 17-minute film. Those who tuned in saw about three additional minutes of footage, but Fox still didn't show the whole seventeen-minute film. In all, the autopsy sequences were only on the screen for thirteen-and-a-half minutes and, once again, that total included clips that were shown repeatedly.

It was not what you would expect from a major network that thought it was broadcasting a history-making film.

It was, however, what you would expect from a network trying very hard not to spoil an illusion.

17. HOW TO MAKE AN "ALIEN" FOR AUTOPSY

Trey Stokes

Much attention has been given to the "alien autopsy" film footage used in the "Alien Autopsy: Fact or Fiction?" program shown last summer on the Fox television network. (The black-and-white film footage was supposedly of a 1947 autopsy of one of three or four "aliens" whose spaceship was said to have crashed then near Roswell, New Mexico. See *Skeptical Inquirer*, November–December 1995; chapters 15 and 16 in this book.] Many people who saw the program seem to think that professional creature FX (FX is Hollywood jargon for "effects," as in "special effects") artists don't know how this could be faked. I happen to be a professional creature FX artist, so let's have a look at that particular claim.

SPECIAL EFFECTS—THE FINE ART OF FOOLING PEOPLE

The job of a special effects artist involves: (1) Creating stuff in an attempt to fool an audience; (2) Looking at stuff other people created and trying to figure out how they did it; and (3) thinking about how we might have done that other stuff.

My opinion of the "alien autopsy"? Everything I saw in the film could have been done with modern makeup FX techniques. Many of these techniques did not exist in 1947, but my belief is that neither did this film. No theater in 1947 would have shown a film as graphic and grotesque as this, even as part of a Hollywood science fiction movie. Why

---

This article originally appeared in the *Skeptical Inquirer* 20 (January/February 1996). Reprinted with permission.

would someone make a hoax the public would never see? As for another often-heard claim that this "alien corpse," if phony, would have to be the best creature effect ever put on film, well, not only do I think it's a fake, I think it could have been a much better fake.

And I, pardon the expression, am not alone. The FX artist seen on that program wasn't the only award-winning creature designer interviewed. A colleague of mine was also asked to review the footage for the program. He pronounced it bogus. For some reason his interview wasn't used. Since the broadcast, I've spoken to many other people who do this sort of work for a living. I have yet to find one who thinks the "alien autopsy" is anything other than a special effect.

Want to know how to do it? Okay, just don't tell anyone else. These are trade secrets.

A HYPOTHETICAL EXAMPLE

Let's suppose I was asked to create an "alien corpse" for an autopsy scene in a movie. Let's also suppose my client doesn't want to do the "ultimate" autopsy scene—just something that will be acceptable. According to the script, the movie scene will go like this: (1) This is a period piece intended to look like a forties-era documentary; (2) The body is supposed to resemble the commonly accepted "alien" description; (3) The body will be cut open and handled by the actors; and (4) We want to show non-human internal organs.

None of these requirements is especially difficult. I take the job. Once the check clears, I assemble my creative team. Right away, we have some important choices to make. There are two basic techniques we could use to create the original form of our corpse: sculpt the whole thing in clay, or do a body cast. Since we're doing a humanoid character, we might choose a body cast for this job. Once we have our body cast, we can adjust it in various ways to make it more "alien." Even with our adjustments, we'll still be stuck with a mostly human-looking corpse, but the body-cast method is both easier and faster than sculpting the entire alien from scratch.

The body-cast process is essentially this: We get a live human of the approximate size we need and cover him or her with alginate, an organic product that goes on like a paste but quickly solidifies into a rubbery semisolid. (You may be familiar with the stuff—dentists use it to take tooth casts.) We reinforce the alginate with layers of plaster bandage. When we remove the hardened bandages and alginate in two big sections (front and back), we've got a "negative" of our human's body. This will be the starting point for creating our alien corpse. (Many FX companies store body casts from past projects. If we happen to have an existing body cast that fits our requirements we might skip this entire step. Now that's economical!)

If we were really in a hurry we might make our final "alien" from the body cast as is; but that could lead to cosmetic problems later. The better technique is to heat up a big batch of oil clay until it becomes liquid, pour the clay into our mold, and let it cool. Pressing cold clay directly into the mold is another option, too. When we open the mold, we have an instant "sculpture," which we can resculpt until our body is exactly the way we want it. This does require us to make another mold of the finished sculpture, but the improved results will make it worth our while. The end result of our body cast is that the "alien" will have nice muscle definition and all the subtle curves and shapes of a real body.

However, we made one mistake. (Actually we try not to make this mistake—but this is hypothetical, remember?) We cast our human standing up because it was easier to get plaster bandages around the body that way. We forgot our corpse would eventually be seen on its back.

Unfortunately, our finished body won't have real muscles under real skin, so it won't shift and react to gravity like a real body would. This is a chubby little alien we're making—if it were real, the underside of the body would lie flatter against the table. Someone looking very closely might also notice the way the flesh appears to hang sideways, toward the toes, rather than downward toward the table. And because our body-cast subject was alive, the leg muscles will be visibly tensed rather than slack, as a dead person's would be.

Oh well, it's good enough. Let's move on.

CREEPY ALIEN BITS

We need to give our little "alien" friend six fingers and toes—just about the easiest possible way to take a human body and make it appear less human.

We probably didn't get very good copies of the hands and feet from our original body cast—we were trying to get the entire body shape rather than little details like that. It's possible we didn't even include the hands and feet in our original body cast since we knew we'd be replacing them later. Also, our body cast subject was standing—if we did use the original foot position, the feet would be at right angles to the legs. We can't have that—our "alien" will look like a department store mannequin that was knocked over.

So, we make hand and foot casts of our original subject, or anyone else whose extremities are approximately the same size. We use our clay-pour technique again to get instant hand and foot sculptures that we resculpt just a bit, adding the extra fingers and toes. (An equally acceptable method would be to sculpt new hands and feet from scratch.)

We take our finished clay extremities and attach them to our clay body, taking care to position the feet in a relaxed pose. We smooth the surface of the clay over the connections, and our "alien" body sculpture is ready.

"ALIEN" HEADS AND YOU

When it comes to making our "alien" head we have the same options as we did with our body: free-sculpture versus a resculpted cast of a human head. Again, it's a mostly human look we're going for here, so we might start with a person's head cast. Then again we might not. It really doesn't matter either way—creating creature heads is done every day in the FX biz, whether it's to create a makeup we apply to an actor, or to create a dummy head. There are many ways to go about it, depending on the artist's preference.

To go with our chubby little body, we'd probably make a chubby little head with a double chin and bags under the eyes.

And we hope this time we won't forget our "alien" will be seen on its back.

We attach the head sculpture to the rest of the body. Now our entire "alien" sculpture is finished, with the clay skin textured throughout.

TIME FOR THE SECOND MOLD

Because our clay model won't twitch, or breathe, or get claustrophobic, or ask to go to the bathroom, we can make a much better mold of it than we could of our original human subject. In any places where two sections of the mold come together, our "alien" body will show a seam line that will need cleaning up later. But we can be careful to construct our mold with close-fitting joints and put them in places where seams are less likely to be seen on camera.

We'd probably also use silicone rubber as the first layer of our new mold. Silicone will mirror the body's shapes and textures like alginate, but silicone won't dry and shrink like alginate does. Our new mold will last for as long as we need it, and we can refill it to make as many "alien" corpses as we want.

A TRICKY DECISION

Our mold will give us an "alien" that looks good on the outside; but it has to look good on the inside, too. Otherwise we could fill the mold with plaster and start an "alien" lawn-statue business.

We want a thick, wet-looking skin, lots of blood and body fluids, and a set of internal organs. And this isn't a still-photo shoot, it's a movie—so we'd like our "alien" to move in a realistic manner as well. (Yes, it's supposedly dead, but it would be nice if our "examiners" could move it around.)

It's not difficult to build a creature that moves well. It's not difficult to build a creature that can be autopsied. It is difficult to build a single creature that can do both. If we design our creature with movement as our main goal, the required mechanical understructure won't leave much room for the internal organs. If we design it with the autopsy in mind, its thick skin and lack of skeletal structure will prevent it from moving very well. Huge, obvious wrinkles will appear at the joints if our actors try to move the limbs on camera.

Well, it's an autopsy movie, which effectively makes our choice for us. But it's also part of our job to work with directors and actors to help show off our effects to their best advantage. Later, on the set, we'll do our best to obscure the fact our "alien" doesn't move. (If we really wanted to do a classy job we might use our mold to build two identical bodies—one to move and one to autopsy. We'd use the first in the preliminary scenes and let the "doctors" handle it all they liked, then swap the autopsy version for the later scenes. Maybe next time.)

THICK-SKINNED "ALIEN"

We need space inside our "alien" for the abdominal organs and brain. To do this we suspend a "core" inside our mold: a plaster blob shaped to fit neatly inside the torso and head. We place our core to allow the proper amount of air space between it and the interior surface of the mold. When we fill the mold, this air space will become the skin we cut through to get to the organs. (We don't need a core for the arms and legs; we'll just let them fill up with our skin material since we've already decided not to bother making them posable.)

We have several options for skin material. A silicone or gelatin mix will give our "alien's" skin a nice "fleshy" quality, if we don't mind the added expense and complexity. Foam latex, a special mixture that expands to form foam rubber, would give us a body that is soft and spongy with a semirealistic simulation of real flesh. Somewhat tricky to use, it also requires an oven large enough to bake our entire mold overnight. Polyfoam, a self-rising urethane similar to foam latex but less expensive and with no overnight baking required, is quick and cheap. And our "alien" will look quick and cheap, too, unless we're very careful!

All of these are workable solutions, subject to our budget, deadline, and personal preference. If the budget allows, we'd probably spring for the silicone skin: It cuts well, looks real, and paints easily.

Just before we close the mold and inject our chosen filling, we may want to add a bit of structure to certain areas. For example, we could embed some wire into the fingers to make them posable. (This will mean we can't move the fingers on camera.) We inject our mold with our chosen filling and wait for it to set.

OPENING THE MOLD

We open the mold—voilà! An "alien." We cut into the "alien corpse's" back and remove the core, leaving a hollow space for our "alien" guts.

The head requires a little extra attention because we want to peel back the skin and reveal a skull. We're not really going to see very much of this in our final film so all we need to do is put a solid shape—most likely made of plaster or fiberglass—into the hollow left behind by the core. In fact, we'd probably use a duplicate of our core to ensure an exact fit. If we had a bigger budget we might go as far as an articulated underskull with a hinged jaw and eyelids and so on, so our "examiner" could fiddle with the eyes and mouth during the examination. Maybe next time.

We trim and patch the body's seam lines where needed. We give our little friend a quick paint job—it doesn't have to be very detailed because we already know this dummy will only be seen in grainy black-and-white.

We stick oversized eyes in our head and put some sort of film over them. We'll be removing this covering as part of our "autopsy." It won't make a lot of sense, but it'll be icky!

We're done.

Oh, almost done. Let's tear out some of the material on the right thigh, paint a bit of blood on it, and create a big, ugly "wound." Should take an extra half hour or so.

FINAL TOUCHES

We bring our body to the set. Just before filming, we reach through the opening in the corpse's back and paint the interior with blood and goo. Then we put our internal organs into place. Maybe we made some beforehand, maybe we bought some livers and kidneys at the market, or both. We seal the opening (it doesn't have to be a cosmetically perfect job—we'll never see the "alien's" back!) and roll the body over. A few drops of glycerin to make her eyes realistically moist, and she's ready!

Our human actors are ready, too, but first we have to give them some coaching. Because of the way we built our creature, they can't move it at all. They shouldn't attempt to raise the arms or legs, rotate the head, or shift the body. In fact, they can only touch it in the most delicate manner or it will become obvious the "flesh" is nearly solid and not semiliquid like real flesh.

Okay, our actors are up to speed now. Let's shoot this thing.

Roll 'em!

First, we get our "establishing" shots of our critter. We have our actors move around, look at the dummy, point to it, and nod. Then we get a few shots in which they pantomime handling the creature. If they do it correctly, it won't be obvious they're barely touching it. (Not many people are aware of the way real bodies in real autopsies are twisted, turned, and flopped this way and that, so they won't realize how bizarre this "examiner" behavior is.)

While we're at it, we'll try a few close-ups where our actors very carefully move the leg and the hand slightly by gripping them firmly and moving them very slightly, just to the point where the skin would start to fold and wrinkle.

NOW WE OPEN HER UP

Now it's time to cut into the body. Here we employ one of the oldest tricks in the book. We take our scalpel and attach a small tube to the side facing away from the camera. As the actor pulls the scalpel along the dummy, we pump a bit of blood through the tube. The scalpel leaves a line of fresh blood. And if some of the blood we put inside the body leaks through the cut, that's even better!

Our next step is to pull back the skin and reveal the abdominal cavity. But first, a brief pause. Until we open the chest, we can't be sure our body interior looks properly realistic. So we tell everyone to take a break while we open the skin of the chest and "dress" the interior—adding any needed blood or details. Then we bring our actors back in and film them as they pantomime peeling the prepared skin with their cutting tools.

This leaves us with a "missing" scene between the original incision and the skin peeling already in progress. But it's a minor omission—and it covers a multitude of possible sins.

After our skin-peeling scene, we can arrange our organs as needed before we roll the camera again.

Our shaky, soft-focus cinematography should help hide the fact that we're looking at a random pile of disconnected organs. Now we can get loads of film of our actors as they remove these "organs" one by one.

LET'S SEE SOME BRAINS

Now for our big finish we'll cut the "skull" open. We didn't spend a lot of time on our skull, but we'll do this in short takes from various bad angles so there's plenty of opportunity to adjust things as we go.

First, we use our blood-tube scalpel on the scalp. We cheat just a bit and skip the moment where the skull is first exposed to allow for any needed touchup work, then let our actors peel the scalp back. We give our actor a saw and let him grind away on the underskull for a while.

Skipping the actual removal of the skull cap, we shoot the removal of the brain from a low angle where the skull can't be seen. We throw one of our organs in there and roll camera as the organ oozes out.

And that's our big finish. Any questions?

Are you sure that's how the "autopsy" was done? Pretty sure. If not precisely the way I've described it, then something close to it.

Does this prove the film is a fake? Well, no. Although there isn't a single moment that doesn't appear to be faked, it's possible the film is genuine and all the flaws can be explained.

Which of the following is a more plausible scenario? (1) This film depicts an actual autopsy of a real "alien" whose body is constructed so exactly like a Hollywood-style creature effect that professional creature FX artists can't tell the difference; and the film itself happens to have been filmed in exactly the way a Hollywood-style scene would be shot, accidentally omitting dozens of details that would have made the film far more believable—or, (2) This film depicts a staged autopsy of a Hollywood-style creature effect.

Until better evidence comes along, I'm choosing the second option.

18. A SURGEON'S VIEW OF THE "ALIEN AUTOPSY"

Joseph A. Bauer

The remarkable aspect of the alleged Roswell alien-saucer crash is that in nearly fifty years of tenacious efforts to legitimize the event by scores of believers and supposed witnesses and participants, not a single, solitary bit of tangible, credible evidence has been found to support such a fantastic and significant event. Despite the reports of extensive debris found in the field at this alleged crash site; despite the many who allegedly handled material fragments with amazing qualities; despite hearsay that the alien bodies and craft were spirited away with unheard-of government efficiency and conspiratorial secrecy to locations that remain mysterious and unproven; despite all these exceedingly unlikely occurrences, no one has surfaced with a hint of convincing, supportive evidence; not even a tiny piece of that mysterious material scattered so widely and handled by so many has surfaced for examination. Didn't any-one slip a fragment into his or her pocket? And now, perhaps to mark the event's upcoming fiftieth anniversary, someone is apparently trying again to prove this was really an extraterrestrial event—this time with an alien autopsy film.

I recognize that it is far easier to create a hoax than to unmask one. But the question "Why?" effectively *exposes the bizarre scenarios depicted* in the autopsy film *as blatant fabrications.*

Why introduce a film now, when alleged mortal fear of repercussions from the government supposedly silenced all witnesses for decades? If the film is authentic, why didn't someone cash in on it in a big way,

---

This article originally appeared in the *Skeptical Inquirer* 20 (January/February 1996). Reprinted with permission.

decades ago, selling it to the highest bidder in a worldwide auction by an agent assuring anonymity of the source? Other than placing a period clock and telephone in the scene, why didn't the filmmaker use some rudimentary special effects to give the autopsy scenario at least the appearance of being *more* than the *clumsy* gropings of veiled, amateur actors impersonating medical investigators?

Considering that an alien autopsy would have been a unique event, the maker of this film should have attempted at least to give the appearance of the event *being* authentic and credible. Why not use a group of actors trained in instrument handling? Why not progress through a systematic autopsy process, rather than just slash and cut out viscera? And wouldn't it have been better to show the need to take many days or weeks to unravel and comprehend the allegedly unrecognizable, misplaced internal organs? But none of these essential procedures was observed, indicating that the autopsy was not authentic, but was contrived by low budget, poorly advised nonprofessionals.

There was no systematic progression of the autopsy, starting from a careful examination *and* penetration of organs and orifices, particularly since alien lore predicates extraordinary eyes, lack of ears or hearing, imperforated oral cavities and questionable need for gastrointestinal tracts, and no genital or anal structures. Next, skilled unroofing of the body cavities would have been followed by surgically precise and detailed dissection, delineating interrelationships, continuity, and formations of the various unknown internal organ systems, during which time decomposition of the body would need to be prevented by some preservation or embalming process. Indeed, there might have been a rare—no, *unprecedented* and *unparalleled*—opportunity to study an alien corpse; but it was not an autopsy that was needed, but rather, a systematic, lengthy, detailed, precise, anatomic dissection and microscopic study of a well-preserved body by a team of specialists of the various, presumably strange, organ systems. No less than that was done in the initial evaluations of the newly discovered Coelocanths. (When a carcass of this primitive fish, thought to be extinct, was first dredged from the depths of the Indian Ocean off Madagascar, ichthyologists worldwide were involved in its dissection, study, and preservation.)

Instead, the dramatic and graphic autopsy—performed with far less diligence and skill than a routine autopsy—was staged by the filmmaker in two scenes. First, the anonymous, hooded figures stand around ineptly trying to occupy their hands, clearly devoid of the rudimentary skills of manual examination of a body, generally expected of any physician, clinical pathologist, or other medical professional. This is followed by tentative, insecure incising, with the operator's face peering down close to the body from which he or she wants to be shielded by wearing the protective suit. Scene two shows the body open; the same inexperienced, unskilled hands are groping around randomly and unsystematically, and without efforts to recognize or analyze organ structures, relationships, or continuity. The bizarre body contents are blindly chopped out and tossed into pans. Ironically, since the external body structure appears so humanlike, the real question is, why should these internal organs be so unrecognizable?

An autopsy is done to determine a disease process, a deviation from the norm, or the cause of death. When the norm is unknown, as would be the case with an alien body, then a careful anatomic dissection is needed with frequent samples being taken for microscopic examination. Anatomical dissection consists of precise steps of delineation, tracing the continuity and relationship of each fold, loop, or bulge to adjacent structures, particularly if the anatomy is unknown and unrecognized as claimed here.

This poorly performed autopsy may have botched a golden opportunity to learn much about this corpse. But it is consistent with an ill-designed hoax. Observation of how ineptly the instruments are held and used is also revealing, and distinguishes a skilled medical professional from an actor. Scissors, for example, are not held with the forefinger and thumb *awkwardly pointing off sideways,* as was done in the film. Instead, the ring finger and thumb are placed in the scissors' holes, the *middle finger stabilizes, and the index finger is used to direct the scissor tip precisely.* Dissection should be done with judicious irrigation and sponging of obscuring fluids (none was seen in the film); dissection is done with direct vision of the knife or scissor points and not by blindly cutting, as depicted. The chopping out and removal of body contents would have totally distorted the functional and structural relationships of organs and destroyed the functional anatomy.

The peculiar headgear of these hooded operators is also enigmatic. Presumably, the hoods were intended to protect against microbes, vapors, or other alien toxins. But as shown, the hoods would cause rapid asphyxia from anoxia and accumulation of exhaled carbon dioxide. Where are the pumps and hoses necessary to supply fresh air to the operators? Without a circulating air supply, the visors would also have become rapidly fogged by condensation, and vision would be obscured. The lack of a detectable air supply suggests that the hoods used for this film were sufficiently porous for air exchange to occur freely, and thus would provide no protection against toxic gases or microbial contagion. All these observations are also most consistent with an ill-designed theatrical mock-up, rather than an actual autopsy of a potentially contagious, decomposing, alien corpse.

The mode of photographic documentation also raises countless questions: Why did a professional photographer repeatedly, if not intentionally, go out of focus and usually position himself or herself behind the actors to obscure the view at the most crucial moments—such as when the cranium (head) was opened? Why was the removal of the skullcap not seen, nor the *in situ* appearance of the brain? Why was a movie camera chosen for documentation (since movie cameras were known to have a focus problem) when efficient 35 mm still cameras with close-up lenses and color film were available at the time and commonly used for medical/surgical/pathological documentation? Furthermore, why was the camera operator allowed to take away and keep a film, when, according to testimony presented, an otherwise high level of secrecy was exercised and enforced with mortal threats? Why did the camera operator not ship this roll back to the military, as he or she did with the other rolls of film, instead of notifying the military to pick it up; and why did the military—incredibly—allow the camera operator to keep this top secret film? Of course a movie camera poorly focused and poorly positioned would be the choice of someone intending to tantalize, mislead, and not reveal any information in the course of hoax.

Only two conclusions are possible from this film: Either this is the work of beginners attempting to create a hoax to resuscitate the corpse of Roswell crash lore; or, if the film is intended to portray an actual autopsy of an unusual humanoid body (a proposition untenable and *entirely* unsubstantiated), then it is a documentation of the crime of the millennium—the brutal butchery, devastation, and destruction of unique evidence and an unparalleled opportunity to gain some understanding about this deformed creature, regardless of its origin.

I hope that this critique will not guide someone to produce a more believable alien autopsy.

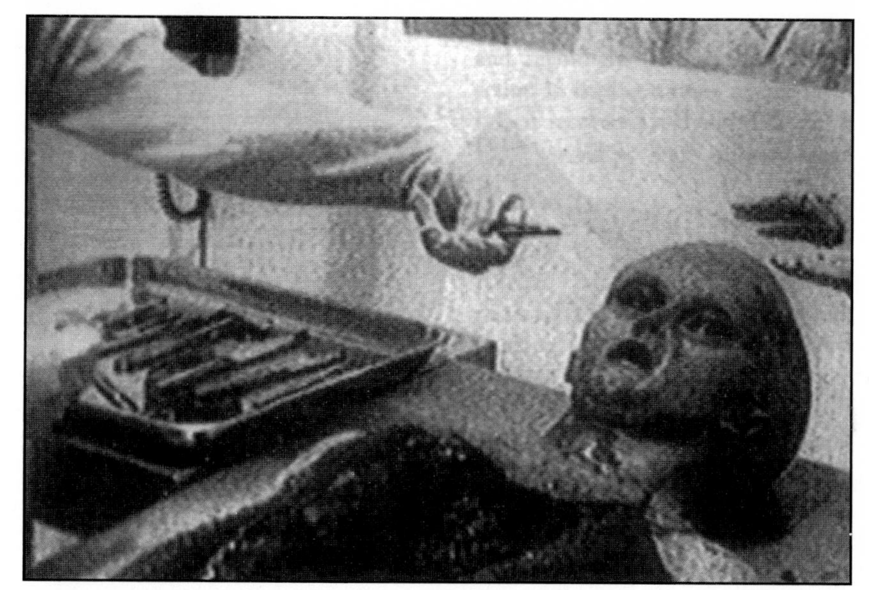

A segment of the reputed "alien autopsy" film, allegedly shot in 1947 and supposed to lepict one of the humanoid extraterrestrials recovered from a crashed flying saucer. Shown on the Fox television program, "Alien Autopsy: Fact or Fiction," the film is an obvious hoax. (From the *Skeptical Inquirer*.)

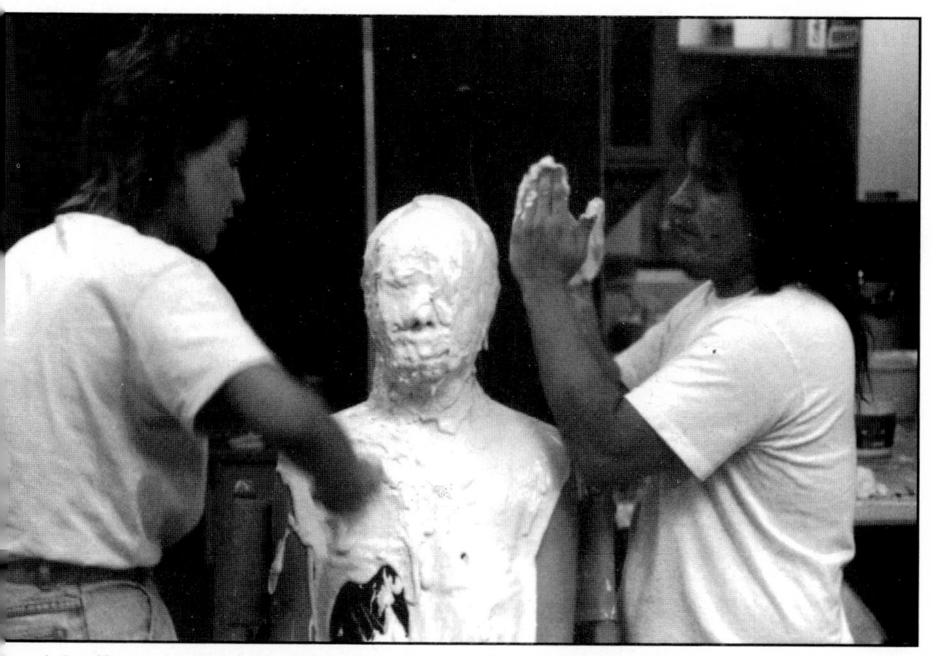

pecial effects technicians demonstrate one of the essential steps in producing a fake xtraterrestrial creature like the one shown in Fox's "Alien Autopsy: Fact or Fiction." Iere, alginate is being applied as the first layer of a head cast. (From the *Skeptical nquirer*. Source: Rick Lazzarini/The Character Shop.)

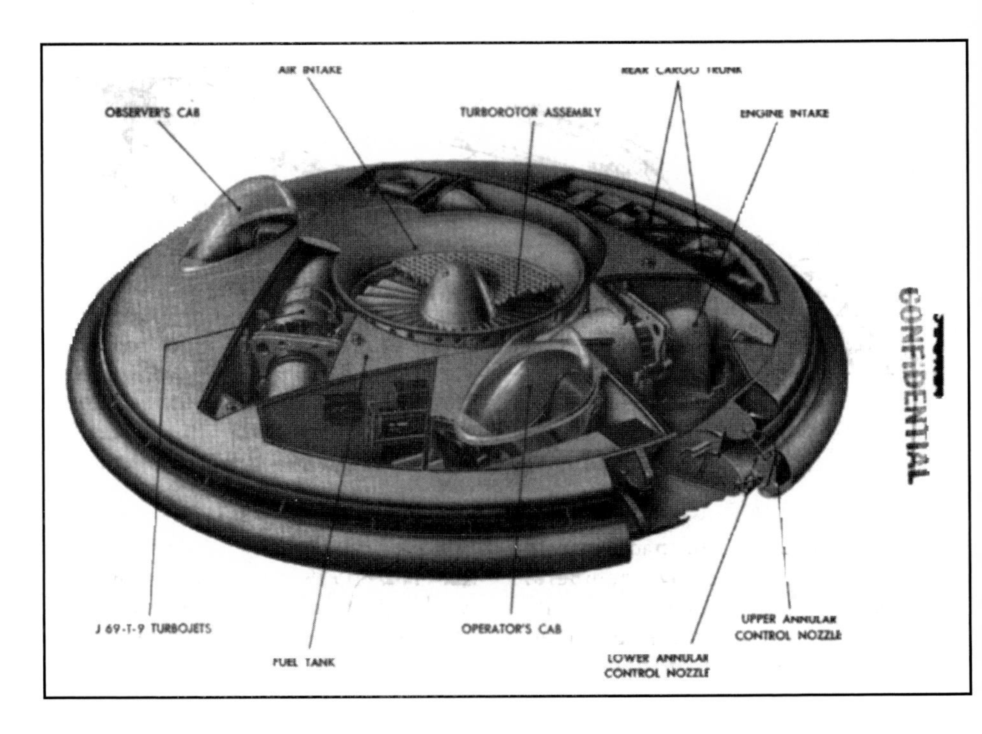

The once-secret Avrocar Test Vehicle. ***Top:*** Cutaway drawing of the 18-foot-diameter disc, showing separate cockpits for its two crew members. ***Bottom:*** A formerly classified "secret" photograph of the Avrocar. Its first untethered flight, November 12, 1959, revealed serious instability problems that were never solved. (From the *Skeptical Inquirer*.)

tmospheric physicist Charles B. Moore
splays a radar reflector similar to those
rried aloft on trains of balloons in
roject Mogul experiments he helped
unch from Alamogordo Army Air Field
New Mexico in June and early July
)47. New York University Flight #4 cared three of these reflectors and before
ing lost was tracked within 17 miles of
e spot where rancher Mac Brazel later
covered debris that prompted the
mous "Roswell incident" case. (From
e *Skeptical Inquirer*. Photo: Dave
iomas.)

A crop circle of the "double pictogram" type that appeared in East Kennett, England. (From the *Skeptical Inquirer*. Photo by Busty Taylor/Centre for Crop Circle Studies.)

A project site of the Search for Extraterrestrial Intelligence (SETI) in Argentina, sponsore by the Planetary Society. (From the *Skeptical Inquirer*. Photograph courtesy of th Argentine Institute of Radioastronomy and the Planetary Society.)

Part Four
OTHER UFO CASES

19. THE GREAT EAST COAST UFO OF AUGUST 1986

James E. Oberg

At about 10 P.M., EST, on Tuesday, August 12, 1986 (0200 UT, August 13), nearly the entire eastern half of the United States was treated to a spectacular celestial apparition. Millions of people were outside looking for Perseid meteors, and many of them had their astronomical instruments and cameras at the ready. So when a bright cloudlike UFO (for it was a genuine unidentified flying object, at least for a day) appeared in the eastern sky, moving from right to left, it had probably the largest audience of any UFO ever witnessed in North America. Sightings occurred from Georgia (Florida was socked in with clouds) to Louisiana to Houston, Texas, to Tulsa and Oklahoma City, to Illinois, Kentucky, Michigan, Ontario and Quebec, and all points in between: South Carolina, Virginia, Massachusetts—the whole eastern seaboard.

Descriptions of the object and its motion varied, but a general picture soon emerged. It was called, in turn, a pinpoint, a moving spiral, a glowing cloud, and a big ball of fire. In Houston, Don Stockbauer described an orangish nebulosity surrounded by an irregularly shaped white cloud elongated vertically, with a dim starlike nucleus. Brenda Newton of Rochester, New York, recalled: "It started to get bigger and it had a tail. By the time we got out of the truck, it had begun to spiral. It lasted for a few minutes, then became like a dim star and floated toward the west." The vice president of the Syracusan Astronomical Society (New York) said it resembled a "reflection of the moon off a cloud, but it was very iridescent, very vivid." Wayne Madea, an amateur astronomer in northern

This article originally appeared in the *Skeptical Inquirer* 11 (Winter 1986–87). Reprinted with permission.

Maine, saw a bright starlike object emit a luminous, rapidly expanding donut-shaped cloud; through a telescope Madea saw "a pinpoint of light, like a satellite, traveling with the cloud."

As it turned out, amateur radio listeners—"hams"—were also receiving radio signals from space at that very moment. And that was the key that led many independent observers to solve the apparition quickly and accurately. Others did it the hard way, as I did, with the application of basic principles of spaceflight and orbital motion.

My involvement began at noon, August 13, when, at a "brown bag" luncheon meeting of astronomy enthusiasts, a report of a bright light in the eastern sky, seen from Houston, was discussed. Returning from lunch, I received a phone call from David L. Chandler, a writer for the *Boston Globe* with whom I had discussed other spaceflight stories months before. He filled me in on the sightings, and I suggested he check about space launchings, particularly the Japanese launch (which I had also learned of that morning). At first there was difficulty in ascertaining the exact launch time (International Date Line, and confusion at the Japanese representative's office over EST vs. EDT), but an hour later, armed with a good liftoff time and with known orbital inclination and period, I was able to produce a hand-calculated map that showed the object heading up the East Coast at about 10 P.M. EST. Its altitude was about 1,500 km (almost a thousand miles), quite high enough for it still to have been sunlit even though the ground below had been dark for more than an hour.

Part of my advantage was a long familiarity with similar apparitions caused by space launches elsewhere in the world, most notably over South America (Soviet launchings from Plesetsk) and Australia (American launchings from Cape Canaveral). So my initial hypothesis about a propellant venting sprang quickly to mind. Such a phenomenon was unheard of over North America, but the new Japanese rocket test was also the first of its kind.

The names of the vehicles involved were a little confusing. The booster was called the "H-1," and it was the first launch. Its second stage was powered by the new "LE5" engine, using super-cold liquid hydrogen as fuel. Two payloads were deployed: an amateur radio satellite variously called JAS-1 (Japanese Amateur Satellite #1), JO-12 (Japan OSCAR 12), or "Fuji" (by the builders); the geodetic mirror satellite, EGP ("Experimental Geodetic Payload"), or "Ajisai" ("Hydrangea Flower"). The booster was launched from Tanegashima Island off Kyushu at 5:45 A.M., JST, August 13 (2045 UT, August 12), after a fourteen-minute hold. Precise tracking data from NORAD allowed a perfect match of sightings to space vehicle.

I then reported my results to the Smithsonian Scientific Event Network in Washington, D.C.; to NORAD Public Affairs in Colorado Springs; to the Committee for the Scientific Investigation of Claims of the Paranormal (which got me on a Buffalo, New York, radio show that had aired live accounts of the UFO on Tuesday evening); to NASA Public Information at the Kennedy Space Center in Florida; and to the MUFON research group in Texas. This in turn resulted in my receiving about twenty additional telephone calls from news media throughout the east.

Chandler's story appeared in the *Boston Globe* (page 6) on Thursday, August 14. It moved over some national news wire, too, since it also appeared in the same day's *Houston Chronicle* and elsewhere. On Friday, I did an interview with an Associated Press reporter from Louisville, Kentucky, and that story moved nationally over the weekend.

Within a week, the "UFO" was stuffed, boxed, and buried. (It should also have been seen from Central America, the Caribbean, and the northern coast of South America—those reports may dribble in over the next few months.) But it was a marvelous experience for the witnesses and for the analysts, and several interesting points can be raised about "UFO reports" based on this fortuitous experiment.

Several interesting events involved coincidences. Caught up in the excitement, Tim Jones, an air traffic controller in Syracuse, reported three different-colored lights randomly moving and hovering for forty-five minutes (but it turned out he was watching B-52s land at nearby Rome AFB, several hours after the real UFO). His account was carried in the nationwide news media, but the solution rarely was. In Clark County, Kentucky, residents were panicked by an explosion while the light show was going on—but the sheriff later got an anonymous phone call confessing to setting off illegal fireworks. Recalled County Deputy Larry Lawson: "The people said their homes shook and windows vibrated as if there had been an explosion or earthquake. . . . They said the whole sky lit up. All these people weren't imagining or seeing things. Some of them were very terrified over it right after it happened." These illustrate the power of coincidence, in which two concurrent independent events can easily (and erroneously) be integrated into a single unsolvable mystery. Also, the emotions (such as fear) of witnesses are no measure of the authenticity of their original perception.

One other amusing aspect was the wide variety of half-baked explanations offered for the "UFO." Some speculation associated it with the meteor shower, a barium cloud, or a satellite burnup, while other guesses associated it with an explosion of the Japanese satellite. Professor Richard Stoner of Bowling Green State University in Ohio was quoted as saying: "It is caused by little bits and pieces of dust from the comet. They're very small, but if there were a larger piece, an icy piece of material, something about the size of a snowflake, it might well cause something like this. It would vaporize and leave a glowing cloud behind it." Astronomy professor Martha Haynes of Cornell didn't trust the observers: "When people who watch the stars once, maybe twice a year go out and look hard for a while, they're bound to see things they think are strange. . . . When you're in that mind-set anything like the light of a plane on the horizon looks strange." John Bosworth of NASA's Goddard Space Center scored a near-miss when he attributed the reports to glints off the EGP satellite's mirrors, reflecting moonlight: "I suspect that's what they saw," he told a reporter. The National Weather Service and the Seattle-based National UFO Reporting Center agreed "it was some sort of natural phenomenon."

A number of people, however, got it right, and right from the start. Tom Bolton of the David Dunlap Observatory north of Toronto told newsmen it was caused by release of something from a satellite: "The satellite was actually seen in the telescope here and we had a report from an amateur astronomer (who) saw it and saw the release of the material from it—but we're not sure which satellite it was and we're not sure what the material was that was released." A number of amateur radio people also told their local papers exactly the true story: For example, Richard C. Eaton of Fayetteville, New York, a retired General Electric engineer, was quoted in the *Syracuse Herald Journal* as suspecting the cloud was part of the Japanese launch.

The spiral form also was intriguing. In Syracuse, amateur astronomer Denise Sabatini reported: "It started out as a pinpoint of light. It was as if it were releasing some type of reflective gas into the air, and as the gas was released into air, it was as if it was spiraling around the pinpoint of light."

The spiral was "like pouring milk into coffee." Astronomer Karl Kamper at the David Dunlap Observatory described the object as starlike surrounded by a small spiral cloud. (He told newsmen the spiral could have been fuel spilling from a damaged satellite and said it must have been extremely high in the sky.) Chuck Barnes, head naturalist at the Troy Farm and Nature Center near Detroit, had been giving an outdoor lecture on meteors when the UFO appeared: "It was glowing like a spiral pinwheel standing on end and moving on a line from southeast to northwest," he told newsmen (the motion was actually from southeast to north*east*); "It appeared to be five or six times larger than a full moon." In Massachusetts, an amateur astronomer watched the plume from the rocket perform two full turns in four minutes, painting the spinning spiral as he watched.

The relevance of these perceptions to other UFO reports is connected with a series of night-time sky spirals seen over China in the late 1970s. While UFO enthusiasts have accepted them uncritically, experienced analysts have voiced the suspicion that they actually involved space launchings (much like the H-1 over America on August 12). These intuitive suspicions were encouraged by a recent official Chinese disclosure of the cancellation of the "Windstorm" space booster, which through the 1970s was being developed in competition with the "Long March 3" booster; there were several flight tests, including one unsuccessful satellite launching, although precise dates were not provided. Further disclosures may allow a precise connection between "Windstorm" space shots and the "spiral UFOs" over China.

Another interesting phenomenon was the way in which UFO groups seemed to get a type of description different from those reported to the national news media. Robert Gribble of the National UFO Reporting Center in Seattle got more than a hundred telephone calls, consistently describing an object shooting straight up into the sky until it mushroomed at a certain altitude. ("It seemed to hold in a certain airspace," he recalled, adding, "I got no reports of it moving across the sky.") Sherman Larson, with the Center for UFO Studies in Illinois, said his group received numerous calls: "In each case, witnesses said an object appeared to have exploded in the sky and then moved into a cloud." In these accounts, subconscious interpretations by the collectors had evidently colored the straightforward, pure perceptions, and without other accounts the stories collected by the UFO groups could well have coagulated into a "true UFO" if the solution had not been published so quickly. This is a long-recognized (but evidently still serious) problem with anecdotal data collection.

All in all, the great cloud UFO of August 12, 1986, was an exciting, illuminating experience, in more ways than one.

20. THE WOODBRIDGE UFO INCIDENT

Ian Ridpath

In December 1980, something remarkable allegedly occurred in Britain outside the U.S. Air Force base at Woodbridge, near Ipswich. News of the event leaked out slowly, finally hitting the headlines in October 1983: "UFO Lands in Suffolk—and That's Official," screamed the front page of the *News of the World,* Britain's best-selling Sunday tabloid.

The story was sensational. It told of a group of American airmen who were confronted one night with an alien spaceship in Rendlesham Forest, which surrounds the air force base. According to the story, the craft came down over the trees and landed in a blinding explosion of light.

The airmen tried to approach the object, but it moved away from them as though under intelligent control. The following day, landing marks were found on the ground, burns were seen on nearby trees, and radiation traces were recorded. There was even talk of aliens aboard the craft, and allegations of a massive coverup. It had all the ingredients of a classic UFO encounter.

The *News of the World*'s informant was a former U.S. airman. He was given the pseudonym Art Wallace, for he claimed that his life would be in danger if he talked. Yet here he was freely giving interviews to newspapers and television.

While his fantastic story might be doubted, it was impossible to shrug off a memo written by the deputy base commander, Lt. Col. Charles I. Halt, to the Ministry of Defence, which was publicly released in the United States under the Freedom of Information Act. Halt's memo,

---

This article originally appeared in the *Skeptical Inquirer* 11 (Fall 1986). Reprinted with permission.

reprinted in full here, is not as sensational as Wallace's story, but it is prime documentary evidence of a type rarely encountered in UFO cases.

UFO researchers in Britain could scarcely believe their luck. The *News of the World* paid £12,000 for the story. A subsequent book about the case, *Sky Crash,* by UFOlogists Brenda Butler, Dot Street, and Jenny Randles, described it as "unique in the annals of UFO history . . . the world's first officially observed, and officially confirmed, UFO landing and contact." Cable News Network made a documentary about the case.

All that evidence, backed up by the word of the U.S. Air Force, could not possibly have a rational explanation. Or could it?

I have my own detective story about the Rendlesham Forest UFO. Soon after the *News of the World* story appeared, I went in search of local opinions about the case. I made contact by telephone with a forester, Vince Thurkettle, who lives within a mile of the alleged UFO landing site. Immediately I was brought down to earth. "I don't know of anyone around here who believes that anything strange happened that night," he told me.

So what did he think the flashing light was in Rendlesham Forest? I was astonished by his reply. "It's the lighthouse," he said.

That lighthouse lies at Orford Ness on the Suffolk coast, five miles from the forest. Thurkettle plotted on a map the direction in which the airmen reported seeing their flashing UFO, and he found that they had been looking straight into the lighthouse beam.

Could this really be the answer? I visited the site with a camera crew from BBC-TV's "Breakfast Time" program. On the way there, the cameraman indicated that he was skeptical about the lighthouse theory. I didn't blame him.

It was past midnight when Thurkettle took us to the site of the alleged landing, and it felt spooky. The area had by now been cleared of trees as part of normal forest operations, but enough pines remained at the edge of the forest to give us a realistic idea of what the airmen saw that night.

Sure enough, the lighthouse beam seemed to hover only a few feet above ground level, because Rendlesham Forest is higher than the coastline. The light seemed to move around as we moved. And it looked close —only a few hundred yards away among the trees. All this matched the airmen's description of the UFO.

The conclusion was clear. Had a real UFO been present as well as the lighthouse, the airmen should have reported seeing two brilliant flashing lights among the trees, not one. But they never mentioned the lighthouse, only a pulsating UFO—not surprisingly, since no one expects to come across a lighthouse beam near ground level in a forest.

So startlingly brilliant was the beam that the TV cameras captured it

Colonel Charles Halt's memo on official American air force notepaper was headed "Unexplained Lights," dated 13 January 1981, and sent to the RAF. It said:

1. Early in the morning of 27 Dec 80 (approximately 0300 L), two USAF security police patrolmen saw unusual lights outside the back gate at RAF Woodbridge. Thinking an aircraft might have crashed or been forced down, they called for permission to go outside the gate to investigate. The on-duty flight chief responded and allowed three patrolmen to proceed on foot. The individuals reported seeing a strange glowing object in the forest. The object was described as being metallic in appearance and triangular in shape, approximately two to three meters across the base and approximately two meters high. It illuminated the entire forest with a white light. The object itself had a pulsing red light on top and a bank(s) of blue lights underneath. The object was hovering or on legs. As the patrolmen approached the object, it maneuvered through the trees and disappeared. At this time the animals on a nearby farm went into a frenzy. The object was briefly sighted approximately an hour later near the back gate.

2. The next day, three depressions 1½″ deep and 7″ in diameter were found where the object had been sighted on the ground. The following night (29 Dec 80) the area was checked for radiation. Beta/gamma readings of 0.1 milliroentgens were recorded with peak readings in the three depressions and near the center of the triangle formed by the depressions. A nearby tree had moderate (.05–.07) readings on the side of the tree toward the depressions.

3. Later in the night a red sun-like light was seen through the trees. It moved about and pulsed. At one point it appeared to throw off glowing particles and then broke into five separate white objects and then disappeared. Immediately thereafter, three star-like objects were noticed in the sky, two objects to the north and one to the south, all of which were about 10 degrees off the horizon. The objects moved rapidly in sharp, angular movements and displayed red, green and blue lights. The objects to the north appeared to be elliptical through an 8-12 power lens. They then turned to full circles. The objects to the north remained in the sky for an hour or more. The object to the south was visible for two or three hours and beamed down a stream of light from time to time. Numerous individuals, including the undersigned, witnessed the activities in paragraphs 2 and 3.

CHARLES I. HALT, Lt Col, USAF
Deputy Base Commander

easily. The formerly skeptical cameraman was convinced. My report was shown the following morning on "Breakfast Time," much to the dismay of the UFO spotters and the *News of the World* reporter.

The lighthouse theory soon had its supporters and its detractors. But there were still too many open questions for the case to be considered solved. For instance, what about those landing marks?

Some weeks later I returned to Rendlesham Forest in search of answers. The landing marks had long since been destroyed when the trees were felled, but I now knew an eyewitness who had seen them: Vince Thurkettle. He recalled for me his disappointment with what he saw.

The three depressions were irregular in shape and did not even form a symmetrical triangle. He recognized them as rabbit diggings, several months old and covered with a layer of fallen pine needles. They lay in an area surrounded by 75-foot-tall pine trees planted 10 to 15 feet apart—scarcely the place to land a 20-foot-wide spacecraft.

The "burn marks" on the trees were axe cuts in the bark, made by the foresters themselves as a sign that the trees were ready to be felled. I saw numerous examples in which the pine resin, bubbling into the cut, gives the impression of a burn.

Additional information came from other eyewitnesses—the local police, called to the scene by the Woodbridge air base. The police officers who visited the site reported that they could see no UFO, only the Orford Ness lighthouse. Like Thurkettle, they attributed the landing marks to animals. The case of a landed spaceship was looking very shaky indeed.

What had made the airmen think that something had crashed into the forest in the first place? I already knew from previous UFO cases that a brilliant meteor, a piece of natural debris from space burning up in the atmosphere, could give such an impression. But I was unable to find any record of such a meteor on the morning of December 27.

Here the police account provided a vital lead by showing that Colonel Halt's memo, written two weeks after the event, had got the date of the sighting wrong. It occurred on December 26, not December 27.

With this corrected date, I telephoned Dr. John Mason, who collects reports of such sightings for the British Astronomical Association. He told me that shortly before 3 A.M. on December 26 an exceptionally brilliant meteor, almost as bright as the full moon, had been seen over southern England. Dr. Mason confirmed that this meteor would have been visible to the airmen at Woodbridge as though something were crashing into the forest nearby. The time of the sighting matched that given in Colonel Halt's memo.

Finally, I turned to the question of the radiation readings. I learned that readings like those given in Colonel Halt's memo would be expected from natural sources of radiation, such as cosmic rays and the earth itself. In short, there was no unusual radiation at the site.

As for the starlike objects mentioned in the final paragraph of Colonel Halt's memo, they were probably just that—stars. Bright celestial objects are the main culprits in UFO sightings and have fooled many experienced observers, including pilots. The object seen by Colonel Halt to the south was almost certainly Sirius, the brightest star in the sky.

If it seems surprising that a colonel in the U.S. Air Force should misidentify a star as a UFO, consider the alternatives. Is it likely that a bright, flashing UFO should hover over southern England for three hours without being spotted by anyone other than a group of excited airmen? And if Colonel Halt really believed that an alien craft had invaded his airspace, why did he not scramble fighters to investigate?

Although UFO hunters will continue to believe that an alien spacecraft landed in Rendlesham Forest that night, I know that the first sighting coincided with the burn-up in the atmosphere of an exceptionally bright meteor and that the airmen who saw the flashing UFO between the pine trees were looking straight at the Orford Ness lighthouse. The rest of the case is a marvelous product of human imagination.

But, somehow, I don't think that my version of the story will make the front page of the *News of the World.*

Postscript. The article above first appeared in the *Guardian,* a respected British daily newspaper, in January 1985. It was written before the U.S. Air Force released a tape recording made by Colonel Halt during his investigations of the "landing marks" and the "radioactivity" in the forest on December 29, 1980. The results of those investigations are referred to in paragraphs 2 and 3 of his memo. I have seen no reason to modify my article in the light of that tape recording.

Since the publication of the article and the release of the Halt tape, Jenny Randles, one of the authors of *Sky Crash* and a leading British UFOlogist, has altered her view of the event significantly. In the November 1985 *MUFON UFO Journal* she writes: "There is nothing on the Halt memo or tape which is inexplicable. Much of it is consistent with the Ian Ridpath lighthouse theory." But she does not accept my explanation. Instead, she now believes that the Halt memo and tape are both part of a coverup for some secret military test or weapons accident.

21. FAA DATA SHEDS NEW LIGHT ON JAL PILOT'S UFO REPORT

Philip J. Klass

The UFO movement, suffering from an extended drought of exciting new UFO incidents to attract media and public interest, got a sorely needed shot in the arm in early January [1987], when it was disclosed that the pilot of a Japan Air Lines 747 cargo airliner had reported an encounter with a giant UFO over Alaska on November 17, while flying to Anchorage from France. The incident had occurred in twilight conditions, starting about 6:15 P.M. local time, with the sun about 11 degrees below the horizon.

According to initial press reports, the incident seemed a classic. The principal witness was an experienced captain, Kenju Terauchi, whose reported visual observations seemingly were confirmed by a USAF/Federal Aviation Administration radar. Additionally, the UFO seemingly paced the JAL 747 for more than forty minutes, offering an extended period for observation by two other crew members of the cargo aircraft loaded with French wine destined for Japan.

Important new insights into the incident have since emerged as a result of the FAA's wise decision to offer a complete data package to the public at modest cost. The available data includes a verbatim transcript of the JAL pilot's tape-recorded radio communications with FAA controllers during the incident, tape recordings and transcripts of FAA interviews with the three JAL crew members in early January, about six weeks after the incident occurred, and a copy of the revealing report that Captain Terauchi submitted to the FAA, also in early January.

This article originally appeared in the *Skeptical Inquirer* 11 (Summer 1987). Reprinted with permission.

In releasing all available data on the incident, the FAA's Alaskan Region public affairs officer, Paul Steucke, noted that his agency "does not have the resources or the Congressional mandate to investigate sightings of unidentified flying objects. We have not tried to determine what the crew of Japan Air Lines flight #1628 saw based on scientific analysis of the stars, planets, magnetic fields, angle of view, etc."

During the initial phase of the November 17 UFO incident, a long-range USAF/ FAA radar sporadically seemed to show a *single* blip in the vicinity of the 747's radar blip—at a time when the pilot was reporting seeing *several* UFOs. Fortunately, the FAA records radar data (for subsequent analysis in event of a mid-air collision or a near-miss), and it was sent to the FAA's technical center near Atlantic City for analysis by radar specialists, to determine if the long-range USAF/FAA radar had indeed detected an unidentified object in the vicinity of the JAL 747.

This analysis showed that the sporadic second blip was due to a phenomenon known as "uncorrelated primary and beacon target," which can occur if the radar energy bouncing off an aircraft does not arrive at precisely the same instant as the signal transmitted back by the aircraft's radar transponder. According to FAA specialist Dennis R. Simantel, who analyzed the data, "these uncorrelated primary returns are not uncommon due to the critical timing associated with the delay adjustments in the aircraft transponder . . . and the target correlation circuitry within the radar equipment."

The FAA data package reveals Terauchi to be a "UFO repeater," with two other UFO sightings prior to November 17, and two more this past January, which normally raises a "caution flag" for experienced UFO investigators. The JAL pilot is convinced that UFOs are extraterrestrial and when describing the light(s) Terauchi often used the terms *spaceship* or *mothership.*

During his January 2 interview with FAA officials, Terauchi said that he believed the "mothership" intentionally positioned itself in the "darkest [easterly] side" of the sky because "I think they did not want to be seen." This enabled the UFO to see the 747 "in front of the sunset and visible for any movement we make." In his report to the FAA, he expressed the hope that "we humans will meet them in the near future."

Terauchi, who was based in Anchorage at the time but has since been transferred back to Japan, noted in his report that his flights over Alaska "generally [are] in the daytime and it is confusing to identify the kind of lights" in darkness. As an example, he described seeing lights from an Alaskan pipeline pumping station reflecting off snow-covered mountains, which initially puzzled him.

(On January 11, a few days after Terauchi gave FAA officials his recollection of the November 17 incident, he again reported spotting unusual lights in roughly the same area while on a repeat flight from Paris to Anchorage. The JAL captain, who has a limited verbal facility in English, asked to record his description of the January 11 UFO in Japanese. Its translation, included in the FAA's data package, resembled Terauchi's description of the UFO initially sighted on November 17: "We see irregular pulsating lights just there is a large black chunk [*sic*] just in front of us. Distance is five miles. It seems to be a spaceship, ah UFO." The pilot reported a similar sighting a few minutes later. But when the USAF/FAA radar failed to confirm the presence of any object, he and the FAA later agreed that these January 11 UFOs were merely lights from small villages being diffused by thin clouds of ice crystals.)

Captain Terauchi, who quickly became an international media celebrity, provided colorful accounts of the incident. But he always failed to mention that two other aircraft in the area that were vectored into the vicinity of the JAL 747 to try to spot the UFO he had been reporting were unable to see any such object. This is revealed in the transcript of radio communications between Terauchi and FAA traffic controllers and their communications with the flight crews of United Airlines flight #69 and a USAF C-130 transport.

United #69 was headed north from Anchorage to Fairbanks at the time that JAL #1628 was headed in the opposite direction along a parallel airway to Anchorage. The FAA asked the United pilot if it could vector him slightly to the left of his intended path, to bring him within several miles of JAL #1628, to see if he could spot and possibly identify the "UFO." As United #69 approached, Terauchi reported the bright light to be at his "nine o'clock" position—roughly broadside and to the left at an estimated distance of about ten miles.

The United captain agreed, and Terauchi was asked to turn his landing lights on briefly to help the United crew locate the JAL airliner. The United crew, looking ahead and to its left, readily spotted JAL, silhouetted against a still faintly light sky, but could not see any luminous object in its vicinity. Shortly before the two aircraft passed, Terauchi was asked again to give the UFO's position, and he reported that it was "just ahead of United"—which would place the bright light to the southeast. Despite the fact that the bright light seemed to Terauchi to be directly ahead of the United jetliner, its crew saw nothing.

In the southeasterly direction, where Terauchi was then looking, was the very bright (–2.6 magnitude) planet Jupiter, which was low in the sky (about 12 degrees) at an azimuth of about 143 degrees relative to true north. From Terauchi's vantage point, Jupiter would appear to be just ahead of United #69. But the bright planet would have been far to the *right* of United's flight path, and its crew would have been looking to their *left* at JAL #1628. Never once did Terauchi report the "UFO's" position relative to a "very bright star," i.e., Jupiter.

Also in the area at the time was a USAF C-130 transport aircraft that was westbound for Elmendorf AFB, flying south of JAL #1628. When the C-130 pilot overheard the FAA communications with JAL, he too offered to try to spot the reported UFO when the USAF aircraft passed near the 747. The USAF crew readily spotted the JAL 747 but reported seeing no other object in its vicinity. The C-130 crew would not have noticed Jupiter, which was to their far *left,* because they were looking at the JAL 747 to their *right.*

While it is commendable that the FAA's Alaskan Region decided to conduct tape-recorded interviews with the three JAL crew members in early January, following inquiries by Japanese news media in late December, in retrospect it is regrettable that the FAA did not think to tape-record discussions with the crew immediately after flight #1628 landed in Anchorage on November 17, when recollections were still fresh. However, when crew members were interviewed separately in January, some significant differences emerged, providing useful insights.

For example, it now appears that the November 17 incident involved two different "types of UFOs," or trigger-mechanisms. As described by flight engineer Yoshio Tsukuba (through an interpreter) during his January 15 interview with FAA officials, the initial UFO was observed for about five to ten minutes at roughly an 11 o'clock position before it disappeared. This is confirmed by the FAA radio communications transcript, which shows the pilot reported the UFO disappearance at 0223:13 GMT, roughly four minutes after it was first reported. Crew members had been observing it for several minutes prior to the initial report.

The second UFO, which Tsukuba characterized as "absolutely different" was visible much further to the left ("nine o'clock") for about thirty or forty minutes. Tsukuba described the initial UFO as a "cluster of lights . . . undulating," which were "different from town lights." Unlike the pilot. Tsukuba said he was unable to describe any particular shape for either UFO. The flight engineer said that, when he was first interviewed by the FAA immediately following the incident, he "was not sure whether the object was a UFO or not. My mind has not changed since then."

During FAA interviews in January, copilot Takanori Tamefuji, who was flying the 747 at the time of the initial sighting, confirmed the flight engineer's recollections that the UFO first sighted was "completely different" from the one later seen further to the left. Tamefuji described what at first appeared to be "two small aircraft" slightly below his own altitude. When the copilot was asked if he could distinguish these lights "as being different" from a star, he replied: "No." (The planet Mars would have been visible to the crew about 19 degrees to the right of Jupiter, but it would not have been nearly as bright.)

When a sketch made by Captain Terauchi, showing a giant walnut-shaped UFO, was shown to the copilot and he was asked if this was what he had seen, he replied: "I don't see anything like this but . . . if we can connect these lights it [would] be a big object, but ah . . ."

There are a number of ambiguities in the report that Captain Terauchi submitted to the FAA on January 2, and in his subsequent interview with an FAA representative, despite the presence of an interpreter. Terauchi generally characterized the initial amber-white lights as resembling the exhaust of jet or rocket engines. In his report, written in Japanese and later translated, 'Terauchi said that a few minutes after first observing the lights ahead and to the left, "most unexpectedly two spaceships stopped in front of our face, shooting off lights. The inside cockpit shined [*sic*] brightly and I felt warm in the face." Neither of the other crew members reported such effects.

All three crew members agreed that the 747's weather radar displayed an echo at a bearing that roughly corresponded to that of the initial lights at a range of about eight miles. The radar display uses color to show the strength of the echo to alert the crew to the potential intensity of thunderstorm turbulence ahead. A red-colored echo indicates an especially strong radar echo and a green color shows the weakest. All three crew members agree that the "UFO blip" was *green.*

This is especially curious if the visual UFO was a giant craft only a few miles ahead, which should have produced an extremely strong (red) return. Flight engineer Tsukuba characterized it as "not a dot, but stream-like." This is confirmed by a sketch drawn by the pilot after landing on November 17. It suggests that the green "blip/stream" was an echo from thin clouds of ice crystals—like those that prompted Terauchi to mistake village lights for UFOs on January 11.

On the night of November 17, there was a nearly full moon that would have been approximately 12 degrees above the horizon at the time of the initial UFO sighting and almost directly behind the JAL 747's direction of flight. This raises the possibility that bright moonlight reflecting off turbulent clouds of ice crystals could have generated the undulating flame-colored lights that Terauchi described.

It would also explain why the undulating lights would periodically and suddenly disappear and then reappear as cloud conditions ahead changed. When the aircraft finally outflew the ice clouds and the initial "UFO" disappeared for good, Terauchi would search the sky for it, spot Jupiter further to the left, and conclude it was the original UFO.

This case is likely to become a classic in the UFO inventory because many people assume that a senior airline captain could never mistake a bright planet or other prosaic object for a UFO. Yet when the late Dr. J. Allen Hynek reanalyzed UFO reports in the USAF files, he found that pilots were as readily misled by prosaic objects as persons in other professions. Numerous air-accident-investigation reports by the National Transportation Safety Board confirm that even *experienced* pilots are not infallible.

I am indebted to astronomers Nick Sanduleak and C. B. Stephenson, of Case Western Reserve University in Cleveland, for their valuable assistance in computing positions and bearings of bright celestial bodies relative to the JAL 747 airliner at the time of the November 17 incident.

22. "OLD SOLVED MYSTERIES": THE KECKSBURG UFO INCIDENT

Robert R. Young

On September 19, 1990, the NBC television network's season opener of "Unsolved Mysteries" featured a half-hour segment on the heretofore little-known "Kecksburg UFO Crash." It was alleged that this involved the crash and recovery by the U.S. military of an unidentified flying object with strange alien markings in the small western Pennsylvania town of Kecksburg, near Pittsburgh, on December 9, 1965.

The program was the tenth most watched in America in a week that saw the introduction of the season's "new" shows. It was viewed in an estimated 17.7 percent of households with television, and on 30 percent of all television sets turned on (*Broadcasting* 1990). Recent surveys for the National Science Foundation report that 2 in 5 adult Americans believe that alien spaceships account for some UFO reports (*Science News* 1986). It therefore seems likely that several million viewers may have been predisposed to accept the premise of the program.

This "saucer crash" has not been widely known to UFOlogists or UFO skeptics because it appears never to have happened. According to a review of all original published accounts, the sole witnesses to the saucer crash apparently were two eight-year-old children who were among thousands in nine states and Canada to view a bolide (brilliant) meteor (Gatty 1965).

Add to this a gullible local flying-saucer buff who has finally found "his own" thrilling flying-saucer crash to investigate; the U.S. Air Force "Project Blue Book" UFO investigating office; "unnamed Pentagon sources"; a secret military satellite launch; the Pennsylvania State Police;

This article originally appeared in the *Skeptical Inquirer* 15 (Spring 1991). Reprinted with permission.

the Kecksburg volunteer fire company; local news reporters who were at first kept away; the twenty-four-year-old recollections of local citizens; and the recent materialization of "new" witnesses.

According to a front-page story in the nearby Greensburg, Pennsylvania, *Tribune-Review* the day after the TV show, some Kecksburg residents, including many observers of the 1965 event and even some portrayed in the program, say it is all a hoax. Some residents blame two local men whose story of a copper-colored 12′ by 7′ "acorn-shaped" object with "hieroglyphic" markings had surfaced only a couple of months earlier—almost a quarter-century after the original publicity.

Tribune-Review staff writer David Darby (1990) reported that more than fifty Kecksburg residents sent a petition to the program's producers in an attempt to stop its airing. The paper reported that these nonbelievers included Ed Myers, the Kecksburg fire chief in 1965, who was portrayed by an actor on the program; Jerome and Valerie Miller, whose home was portrayed as the site of a "military command post" during UFO recovery operations; the owners of the land where the saucer was supposed to have landed; and Kecksburg firemen.

Myers expressed concern. "It's killing me to know this is going nationwide, because there's absolutely no truth to it," he told Darby. "Something's gonna be put in the history books for my grandchildren to read, and it is just not true."

The Millers, the paper reported, deny that their home was a center of military activity. Darby said "whoops of laughter" filled the Miller living room when a group of residents who consider the whole thing a hoax gathered to watch the melodramatic program.

Several elements combined in 1965 to create local hysteria. For several days the world had been fascinated by front-page coverage of the missions of Gemini 6 and 7, two U.S. spacecraft set for a manned joining. The day of the incident (December 9) the *Pittsburgh Press,* widely read in the Keckburg area, reported that Frank Edwards, a nationally known flying-saucer lecturer and broadcaster had arrived in the city to speak. The headline, "Lift UFO Secrecy, Saucer Believer Says," had a "kicker" above it, "U.S. Hush-Up Charged."

However, the *Erie Daily Times* (December 10) reported another event that day that went largely unnoticed: a secret satellite was launched from Vandenberg Air Force Base, California, a launchsite for military polar-orbiting reconnaissance missions. The stage was set.

Shortly after 4:40 P.M. (EST) a brilliant bolide, or "fireball," was seen by thousands in Idaho, Illinois, Indiana, Michigan, New York, Ohio, Pennsylvania, Virginia, West Virginia, and Ontario, Canada, according to reports on December 10 in the *Erie Daily Times*; the *Pittsburgh Press,* the

New York Times, and the *Pittsburgh Post-Gazette.* The fireball was even said to have been seen in California (*Pittsburgh Press,* December 10, 1965). Astronomers from Michigan, Ohio, and Pennsylvania, who had received many reports, concluded the object had been a bright meteor (*Erie Daily Times, Pittsburgh Press, New York Times, Pittsburgh Post-Gazette,* December 10). This was also the conclusion of the Federal Aviation Administration, according to a spokesman at Erie, Pennsylvania (*Erie Daily Times*; *Pittsburgh Post-Gazette*); air force spokesmen in Washington; and unnamed "Pentagon sources" (*Pittsburgh Press, New York Times*).

Reports of bolides are typically inaccurate. Astronomer Frank Drake (1972), after efforts to recover meteorites from fireball reports, has estimated the fraction of eyewitnesses who are wrong about something to be 1 out of 2 after one day, 3 out of 4 after two days, and 9 out of 10 after four days. Witnesses often grossly underestimate the distance of firebans, which may be dozens of miles high. When the meteors disappear over the horizon it is sometimes taken as a "nearby" event (Klass 1974, 42–49).

The 1965 fireball was no exception. It was reported to have "crashed" or "landed" in six widely separated locations. A pilot in the air reported watching as it "plummeted" into Lake Erie (*Pittsburgh Press,* December 10). At Midland, Pennsylvania, west of Pittsburgh, falling debris was reported but police found nothing (*Erie Daily Times, Pittsburgh Press,* December 10). At Elyria, Ohio, west of Cleveland, a woman reported that a fireball the size of a "volley ball" fell into a wooded lot. Firemen reported ten small grass fires but no flying saucer (*Pittsburgh Post-Gazette,* December 10).

At Lapeer, Michigan, forty miles north of Detroit, sheriff's officers investigating the report of "a ball of fire crashing" found only pieces of tinfoil (*Pittsburgh Press,* December 10). The most spectacular report came from Detroit and Windsor, Ontario, where pilots, weather observers, and U.S. Coast Guard personnel reported that a flying object "exploded" over Detroit. Coast Guard boats sent into Lake St. Clair found nothing (*Tribune-Review,* County Edition, December 10). The air force UFO investigating office at Wright-Patterson Air Force Base, Ohio, may have been interested in the recovery of space-launch debris and sent three-man investigating teams from the 662 Radar Squadron, based near Pittsburgh, to Kecksburg and Erie (*Erie Daily Times, Pittsburgh Post-Gazette*).

In Kecksburg the scene had turned into a circus. Little Nevin Kalp had run and told his mother, Mrs. Arnold Kalp of RD 1, Acme, Pennsylvania, that he had seen something "like a star on fire." Going outside she saw "blue smoke" that seemed to come from a nearby woods (Gatty 1965; *Pittsburgh Press,* December 10). Other reports had described a bright trail left in the air by the meteor (*Pittsburgh Post-Gazette,* December 10). A "thump" whose vibration felt by one witness was attributed to dynamiting at a local quarry or to a shock wave heard by many western Pennsylvanians who witnessed the fireball. Mrs. Kalp called a local radio station that had been reporting a plane crash. Soon, according to the *Tribune-Review,* a "massive traffic jam" had engulfed the small town (Gatty, *Tribune-Review,* City Edition, December 10, December 11).

A local volunteer fire policeman informed reporters that the Army and the state police had told them not to let anybody in (Gatty 1965). One result was that an early edition of the Greensburg paper carried a seven-column banner headline atop page one, " 'Unidentified Flying Object' Falls Near Kecksburg," and, "Army ropes off area" (*Greensburg Tribune-Review,* County Edition, December 10).

Captain Joseph Dussia, commander of the Pennsylvania State Police Troop A Headquarters at Greensburg, announced the next day that after an all-night search "absolutely nothing had been found." Reports of something being carried from the area referred only to equipment used in the search, Dussia said. He added, "Someone made a mountain out of a molehill" (*Greensburg Tribune-Review,* City Edition, December 10). The Air Force also announced that nothing had been found (*Pittsburgh Press,* December 10). The next day a *Greensburg Tribune-Review* editorial summarized its staff's independent investigation: Nothing at all seems to have happened (December 11). The official explanations are totally consistent with all published accounts and the present recollections of scores of witnesses.

When does the "unsolved mystery" come in? Now enters Stan Gordon, founder of the Pennsylvania Association for the Study of the Unexplained (PASU), a Greensburg-based group that collects sightings of UFOs, Bigfoot, and other oddities, such as the "Eastern Cougar," an animal that has been extinct for a hundred years. PASU seems to do little research into these events but does issue press releases. Gordon, a thirty-year veteran of saucer chases, is also Pennsylvania director of the Mutual UFO Network (MUFON), the nation's largest surviving flying-saucer group.

Each year in early January PASU issues its annual press release to Pennsylvania newspapers listing exciting reports received during the previous year. Their 1989 release featured an alleged UFO encounter by a Harrisburg, Pennsylvania, policeman (*Latrobe Bulletin,* January 9, 1989). A PASU investigator later said the witness had suffered "severe burns" and a "severe eye injury." MUFON's state director soon turned it into a "returning UFO abductee" encounter, making claims publicly denied by the witness. Local amateur astronomers found the witness had been looking at the planet Venus. The witness refused to be examined by a physician; a PASU investigator "lost" film evidence of the witness's injuries, and a substance Gordon had tested at a laboratory and then described as "strange" and "unusual" turned out to be a common fertilizer (Young 1989).

In 1990 PASU issued a call for anyone with knowledge of the Kecksburg UFO crash to come forward (*Latrobe Bulletin*). With an experienced nose for saucer news, they must have sensed that even after twenty-four years witnesses always seem to be willing to come forward if the case is exciting.

Actually, the Kecksburg UFO tale has been making the rounds among Pennsylvania saucer buffs for some time. Flying-saucer evangelist Robert D. Barry hosts a Saturday midnight program, "ET Monitor," on WGCB-TV, Red Lion, Pennsylvania, a religious station, where he mixes NASA films, UFOria, viewer calls, and occasional Bible readings. Barry mentioned the Kecksburg recovery in a lecture at Elizabethtown College, Elizabethtown, Pennsylvania, on March 22, 1989, and followed on his April 2, 1989, program with the revelation that the incident involved the recovery of "bodies." Later, on his April 23, 1989, broadcast, he stated that no bodies were involved in the UFO accident.

Barry says that years ago he was told by an unnamed NASA informant that the Kecksburg UFO had been tracked, a claim that is contradicted by statements made by a North American Air Defense Command spokesman at the time (*Erie Daily Times*; *Pittsburgh Press*; *Pittsburgh Post-Gazette,* December 10). Barry has also reported, citing Stan Gordon as his source, that a 1965 member of the Kecksburg Fire Company claims it had been contacted by NASA *before* the UFO crashed and asked to keep the public away from the area, a claim contradicted by the original published reports and eyewitness statements (*Tribune-Review,* City Edition, December 10, 1965).

A curious claim, oddly similar to the Kecksburg story, occurred January 28, 1990, on Bob Barry's television program. At 7:10 P.M. (EST) that evening a bright fireball had been seen over much of the East Coast (*Harrisburg Sunday Patriot-News,* January 28, 1990). That night on "ET Monitor" Barry reported that "a Greensburg source," had called to say that "an object landed" nearby at about 7:20 P.M., that the area had been cordoned off, and that the source was "trying to get as close as he could." A well-known baseball philosopher would have been prompted to say that it seemed like "déjà vu all over again."

It is too bad the producers and researchers at "Unsolved Mysteries" didn't scratch around a little. At least fifty folks at Kecksburg could have saved them an embarrassment.

UPDATE

After the 1990 "Unsolved Mysteries" broadcast, more than a hundred people called the show's phone "hotline" to report that they were also witnesses. One man reported details which had been added to the TV show for dramatic effect. Others saw an object like the TV "acorn," even though Kecksburg witnesses said it was long like a rocket. Yet others said that they saw the UFO at Ohio air force bases.

My investigation shows that these Ohio stories bear close resemblances to a 1954 hoax from the same city. In its turn, it had been lifted almost word-for-word from the tale told by two confidence men to Frank Scully, author of the best-selling *Behind the Flying Saucers* (1950), one of the first American flying saucer books.

Kecksburg Crash proponents have accepted all of these new eyewitness claims, even though they may contradict one another. As a result, the Kecksburg incident has now taken on all of the trappings of the classic saucer crash tale, including claims of dead bodies.

No investigator claiming a UFO crash has revealed that photographs of the meteor's cloud train were taken seconds later by two widely spaced Michigan photographers. One of these was published in *Sky & Telescope* magazine's February 1966 issue. The article concluded that the fireball was a meteor. A copy is in the USAF Project Blue Book file in the National Archives. Stan Gordon reported that he has been in possession of this file since 1985.

Michigan State University astronomers Von Del Chamberlain and David J. Krause used the pictures, many eyewitness reports, and a seismometer recording of the sonic boom to estimate the object's speed at about 14.5 kilometers per second. This is well within the speed of meteors entering the Earth's atmosphere and about twice as fast as returning man-made space objects in low orbits.

They determined the meteor's path through the atmosphere. It fell steeply over southeastern Ontario and traveled northeastward, not toward Pennsylvania. A possible orbit for the meteoroid was determined out to the asteroid belt, between the orbits of Mars and Jupiter, where many bright fireball meteors originate.

An article was submitted and published in 1967 in the *Royal Astronomical Society of Canada Journal* (vol. 61, no. 4). On April 21, 1992, I sent a copy of this Chamberlain and Krause article to UFO researcher Gordon. Four and a half years have passed, but I have received no response.

Those who claim to have seriously researched the 1965 event must account for these photographs and published research, but they continue to ignore their existence. One can therefore reasonably conclude that it is the UFO investigators, not the U.S. Government, Kecksburg doubters or the Pennsylvania State Police, who are guilty of a deliberate, continuing coverup of documents and photographs which reveal the truth about the Kecksburg "UFO crash."

REFERENCES

Broadcasting. 1990. (Cites Nielsen and its own research.) P. 40.

Darby, David. 1990. *Greensburg Tribune-Review* (Greensburg, Pa.), December 10, p. 1.

Drake, Frank. 1972. "On the Abilities and Limitations of Witnesses of UFO's and Similar Phenomena." In *UFO's: A Scientific Debate,* edited by Carl Sagan and Thornton Page, 247–57. New York: Cornell University Press and W. W. Norton.

Gatty, Bob. 1965. "Unidentified Flying Object Report Touches Off Probe Near Kecksburg." *Greensburg Tribune-Review,* December 10, p. 1.

Klass, Philip J. 1974. *UFOs Explained.* New York: Random House/Vintage. See pages 42–49.

Science News. 1986. Vol. 129: 118.

Young, Robert R. 1989. "Harrisburg 'UFO Incident' Stimulated by Venus." Unpublished manuscript by the author.

23.
GULF BREEZE UFO CASE
Robert Sheaffer

A great deal has been happening in Gulf Breeze, Florida, the now-undisputed UFO capital of the world, since it was last mentioned in this column (Summer 1989: 363). "Mr. Ed," the once-anonymous photographer around whom alien beings and their craft seemed to swarm, has now surfaced as building contractor Ed Walters, who with his wife, Frances, is coauthor of a successful book, *The Gulf Breeze Sightings: The Most Astounding Multiple Sightings of UFOs in U.S. History.* Despite its title, the book seems to have been singularly unsuccessful in convincing anyone of the authenticity of Walters's Polaroid saucer photos who wasn't convinced already. The book has been attacked bitterly by a number of seasoned, longtime UFO proponents, many of whom are still shaking their heads sadly at the way MUFON—the largest surviving UFO group—has uncritically promoted Walters's yarns. Whatever faults the book may have, it was good enough to earn Walters a $200,000 advance, according to news reports, as well as the sale of TV rights for a planned miniseries that may net as much as $450,000.

In spite of this abundance of wealth and fame, life has not been a bed of roses for Gulf Breeze's most famous citizen. This past June, the *Pensacola News Journal* reported that the man who now lives in the house Walters occupied at the time of the alleged alien blitz had discovered a model flying saucer, apparently forgotten, hidden away under some insulation in the garage attic. Using this model, news photographers were able to create numerous photos of UFOs looking remarkably like those that

---

This article originally appeared in the *Skeptical Inquirer* 15 (Winter and Summer 1991). Reprinted with permission.

thrust Ed Walters into the limelight. Walters, of course, denies all knowledge of the model, even though paper wrapped around it contains part of a house plan he himself drew up. He claims that particular plan was not drawn up until after he vacated the residence, but Gulf Breeze city officials say the plans are two years older than Ed says they are. Walters suggests that the model was probably planted in his former residence by some "debunker" who, he insists, "will do whatever [is] necessary to debunk a case." If so, one would expect a dastardly debunker to leave the incriminating model in a place where it would likely be found, and not so well concealed that it was not discovered until the current resident of the house undertook to modify the cooling system.

But in a development even more potentially damaging to Ed's credibility than the discovery of the model, a Gulf Breeze youth named Tom Smith has confessed that he and two other youths, one of them Ed's son Danny Walters, assisted Ed in the production of hoax saucer photos. Worse yet, Smith has a series of five UFO photos taken with his own camera to substantiate this collaboration. But Walters claims that young Smith took the UFO photos unassisted and that they are genuine. Veteran saucerer James Moseley became outraged when Charles Flannigan, MUFON's Florida State Director, ordered him not to talk to any of these youths until the MUFON hierarchy had a chance to interview them first! Moseley charges MUFON with practicing "damage control" by its insistence on getting to the youths first "in order to plug up any holes in their stories." Nonetheless, Moseley is still impressed by Ed Walters's apparent sincerity and is not convinced that the Gulf Breeze UFO photos were faked.

The frenzied wave of supposed UFO photos and sightings in Gulf Breeze, Florida, touted as "the most astounding multiple sighting of UFOs in U.S. history," continues to fascinate and to entertain, even as it sinks slowly into the sunset. When we last left prime UFO-spotter Ed Walters (*Skeptical Inquirer,* Winter 1991: 135), he was grappling with the fact that a model UFO had been discovered in his former home and with a neighborhood youth's confession that he had helped Ed fake some of the photos. Since then, things have gotten worse.

In the wake of these embarrassing disclosures, MUFON, the largest surviving UFO group (which has always taken a staunch pro-Walters line), asked two of its most respected investigators, Rex and Carol Salisberry, to take another look into the case. At one time MUFON's hierarchy expressed the highest degree of confidence in the Salisberrys' investigative skills and presented them a special award at last year's MUFON conference for outstanding investigations. But when the two reported their findings that Ed Walters was "adept at trick photography" and had faked the photos, MUFON suddenly lost all confidence—not in Walters, of course, but in the Salisberrys—and tossed them out of the organization and disavowed their report. Carol Salisberry writes that the local MUFON group "is out to lynch us, and the hierarchy is goading them on."

Nonetheless, local hotels, restaurants, and travel agents are doing all they can to cash in on the UFOria. The director of tourism for the Pensacola Area Chamber of Commerce called the situation in Gulf Breeze "very, very positive" and said UFO tour groups are being arranged. A local travel agent adds that UFO-seeking tourists have "brought in a lot of money to Gulf Breeze merchants when they desperately needed it." But six Gulf Breeze visitors last year most certainly did *not* benefit from their UFO pilgrimage. Last July, six American soldiers with top-secret clearances went AWOL from their Military Intelligence Brigade in Germany and mysteriously turned up in Gulf Breeze. Arrested by local police, they were taken into military custody at Ft. Banning, Georgia. Newspaper reports suggested that the six were members of a "Rapture" or "end-of-the-world" cult that believed that "Jesus Christ was an astronaut." They were reported to have come to Gulf Breeze to "hunt down the Antichrist." Conflicting accounts were published about the motivation for this incident, and several news organizations received a bizarre letter—it looked as though it had been produced on a teletype—demanding that the U.S. Army "Free the Gulf Breeze Six." Whatever the reason, charges were soon dropped against the soldiers, who were released and then discharged.

If press reports were correct about Walters receiving a $200,000 advance for his recent book, the publisher isn't likely to recoup that investment; and if Walters was paid $450,000, as was reported, for the rights to a television miniseries, nobody seems to be rushing to produce it. Walters was not even successful in his attempt to use his recent fame as a springboard into local politics; running for a seat on the Gulf Breeze City Commission, he came in dead last. Even veteran saucerer James Moseley, who until recently professed being impressed by Walters's apparent sincerity, now concedes "it looks more and more likely that the Walters story is a hoax." Thus the three-ring flying-saucer circus in Gulf Breeze seems to be slowly winding down. Fear not: it does not spell the end of saucerdom. Before long, more astonishing UFO photos will be taken somewhere else by some other enterprising soul, and the UFO show will once again swing into high gear, with a fresh new cast.

24.
THE BIG SUR "UFO": AN IDENTIFIED FLYING OBJECT

Kingston A. George

The air force obtained some unusual photography while experimenting with very sensitive optics equipment during ICBM launches on the West Coast nearly thirty years ago. Three years ago, in an article titled "Deliberate Deception: The Big Sur UFO Filming" (Jacobs 1989), one of the members of the experimental team claimed that the objects observed were beyond normal technical explanation and implied that the government had been communicating with aliens from outer space. Specifically, he claimed that the team had photographed an "intelligently controlled flying device." He asserted that it emitted "a beam of energy," its capabilities were beyond the science and technology of our time, and it was therefore probably "of extraterrestrial origin." He concluded that we had knowingly photographed a "demonstration . . . put on for our benefit for some reason by extraterrestrials." I was the project engineer for these experiments. This article is intended to provide a more rational account of the sightings of September 1964 and to supply firsthand facts that should loosen any attachment the uninformed might have to Bob Jacobs's version.

THE DEPLOYMENT

The United States Air Force conducted a test of a special light-sensitive telescope high up in the coastal mountains in the Los Padres National Forest above Big Sur, California, between August and November 1964. The

This article originally appeared in the *Skeptical Inquirer* 17 (Winter 1993). Reprinted with permission.

objective was to collect low-light-level photography of missile launches into the Air Force Western Test Range from Vandenberg Air Force Base, situated a little over 100 miles to the south. The Big Sur angle presents a unique side-look during test launches, and paper studies convinced some of us that photo data from that location could be of significant value. Local telephoto-lens coverage from Vandenberg AFB is often obscured by the prevailing fog, while the special telescope could be placed at 4,000-feet altitude. Nine of eleven launches from Vandenberg were successfully covered during the three-month deployment (George 1964).

The 24-inch mirror telescope we borrowed was built in the 1950s on a modified 5-inch gun mount by Boston University under government contract. Owned and operated by the Range Measurements Laboratory of the Air Force Eastern Test Range, the B.U. Scope, as we called it, later supplied the television network feed during Saturn rocket launches in the sixties and seventies. It employed one of the most light-sensitive systems of the time, an image orthicon television camera tube.

An image orthicon "sees" stars quite well even in twilight. The brightest ones would bloom on the closed-circuit TV monitor to form a blob, with size related to brightness, and also leave a persistence tail behind as the telescope panned across it. The tracking operators used handwheels to constantly make tiny adjustments, and the TV screen resembled a pool of vigorous tadpoles. Today, a similar modern instrument detects stars several orders of stellar magnitude less bright than the best we could do in 1964.

The project was remarkably successful. Soon after we returned the borrowed instrument, a long-term plan was started for a permanent site. An up-to-date telescope is operated today in the Big Sur area by the Western Test Range's successor, the 30th Space Wing of the Air Force Space Command.

I was the project engineer for the telescope experiment, and Lieutenant Bob Jacobs was one of the key field team members who, it later developed, was technically not authorized to view the pictures we were collecting. Bob was named the on-site commander by the 1369th Photo Squadron and managed the logistics of the operation at the Big Sur location. Years later, for reasons I can't fathom, Bob claims we witnessed an intelligent UFO in action around an Atlas warhead, followed by an Air Force coverup. He provides details of his weird claims in an article for the *MUFON UFO Journal* (Jacobs 1989). What we saw was indeed unique and startling, but it definitely does not require invoking UFOs with purposeful goals and advanced weapons.

THE THREAT TO NATIONAL SECURITY

The immediate success of the 1964 project led to a serious problem: we not only could see and gather data on the missile anomalies as hoped, but we also were viewing details of warhead separation and decoy deployment that were considered by the air force to be highly classified. The air force strives to be quite rigid in its approach to handling classified information, yet there were suddenly dozens of airmen, civilians, and contractors viewing data normally reserved to a few persons with the highest level of clearance. Of course at first no one realized the significance of the data.

By the early 1960s, the USSR had beaten the United States into space and set numerous "firsts," demonstrating an alarming degree of sophistication in rocketry and the space sciences. The limits of what was technically possible in space were not well defined for the military leadership. The United States owned radars that could detect incoming warheads thousands of miles from their targets and anti-missile missiles that could theoretically knock out an incoming reentry vehicle above the atmosphere. Could the Soviets nullify our land- and submarine-launched missiles with an anti-ICBM system? Today we can say it was naive to think either we or the USSR could have fielded much of a defense against ICBMs with the technology available in the sixties. But in 1964, the military leadership had to react as though a defense against the ICBM forces was around the corner.

DAWN ON SEPTEMBER 22, 1964

Just after sundown and just before sunrise, there is a period of time when objects at high altitude overhead are sunlit to an observer who is in darkness on the earth's surface. About fifteen to twenty minutes before dawn, when the sky is quite dark, conditions are poised for optimizing the contrast and range of detection for objects hundreds of miles distant.

Such was the case during an Atlas launch nicknamed "Buzzing Bee" before sunup on September 22, 1964. On the TV screen, we watched the Atlas climb into the sunlight and shed its booster engine section about two minutes after launch. The sustainer engine shut down some two and a half minutes after that, all normal for the Atlas, and we could still see the missile tankage against the dark, starry sky! And then, astonishingly, we saw a momentary puff of an exhaust plume, bright enough to "bloom" on the television monitor, and an object separated from the tank—the reentry vehicle (RV) was released to follow its own trajectory to the target area. This was followed by two smaller puffs that also bloomed on the monitor, and then two groups of three objects became distinct from the sustainer tank and the RV. We watched all the objects slowly grow in separation from one another for another minute and a half. Then the objects grew so dim, and the tracking so erratic, that the operation was halted. We had watched the flight for about eight minutes.

The Atlas was supposed to release decoys, simulated RVs to confuse and overload a missile defense system. The timing of the puffs we had seen was in the right ballpark. Beyond that, we needed expert assistance to help explain the images. We carried a canister containing a thousand feet of 35 mm black-and-white film (at that time, video was recorded by a synchronized film camera viewing a kinescope) to Vandenberg AFB, processed it, and began showing it with some excitement to the Atlas missile development people.

The reaction was startling! Soon after the first showing to the director of operations, all the top brass at Vandenberg had seen it and a copy was being made to fly to HQ Strategic Air Command at Omaha. The classification was quickly changed from Secret to Top Secret. Buzzing Bee had opened an entirely new chapter in ICBM tactical thinking.

JACOBS'S OBSERVATIONS

Jacobs reports in the MUFON article that he witnessed a saucerlike UFO circle the Atlas warhead, then direct a laser beam at it that bumped it out of the way and caused it to tumble out of orbit [*sic*] and miss the intended target by hundreds of miles. There are several fundamental flaws in that statement. To begin with, the Atlas was suborbital, as all ICBMs are, and it did not miss the target.

The image of the warhead, even if viewed exactly side-on, would be less than six-thousandths of an inch long on the image orthicon face, or between two and three scan lines. We could not resolve an image of the warhead under these conditions; what is detected is the specular reflection of sunlight as though caught by a mirror. Practically all the data collected by the B.U. Scope on hard objects was through specular reflection. The same principle is involved in the little hand mirrors provided to military pilots so that an air search can find them by the glint of reflected sunlight if necessary.

We could also see the engine exhaust as a large gaseous plume that dissipated rapidly outside the earth's atmosphere. The small charges that released the decoys were seen as short flashes about as bright as a dim star. Nothing "circled" any of the images.

A laser beam (or any directed-energy beam) is invisible in the vacuum of outer space. We are able to see the path of a laser beam in a surface environment only because of dust particles and ionization in the surrounding atmosphere. A laser beam damages a target not with momentum, but by heating and melting it.

Six conclusions are given by Jacobs in the MUFON article requiring comment.

Jacobs Conclusion 1: "What we photographed that September day in 1964 was a solid, three-dimensional, intelligently controlled flying device." Bob is referring to his impression of something circling the warhead when he says "intelligently controlled." Nothing of the sort happened.

Jacobs Conclusion 2: "It emitted a beam of energy, possibly a plasma beam, at our dummy warhead and caused a malfunction." As noted above, the fact is that energy beams cannot be seen unless they hit something or pass through an atmosphere. We might see a target begin to glow with heat if we were close enough.

Jacobs Conclusion 3: "This 'craft' was not anything of which our science and technology in 1964 was capable. The most probable explanation of the device, therefore, is that it was of extraterrestrial origin." This remark must be Occam's Razor upside-down and backwards! Everything detected was indeed a product of our science and technology, although we had never had a direct view of it before. The Eastern Test Range people who operated the B.U. Scope for us had never seen views like this either, mainly because the telescope was situated to look "up the tail" of the launches on the East Coast. Also, images are seriously degraded by the light passing through a great deal more atmosphere than on our 4,000-foot mountain.

Jacob Conclusion 4: "The flashing strikes of light we recorded on film were not from laser tracking devices. Such devices did not exist then aside from small-scale laboratory models." In 1962 I evaluated the feasibility of using a carbon-dioxide laser to illuminate launch vehicles hundreds of miles away! In the late sixties the Range Measurements Laboratory at the Eastern Test Range operated two high-powered lasers in the visible spectrum for imaging space objects at night on a regular basis. But Bob is correct in saying that the observations in 1964 did not involve lasers—and, I would add, neither intra- nor extraterrestrial.

Jacobs Conclusion 5: "Most probably, the B.U. Telescope was brought out to California specifically to photograph this event which had been prearranged. That is, we had been set up to record an event which someone in our Government knew was going to happen in advance." My supervisor at the time, Gene Clary, and I would have been thrilled to have had any kind of support from anywhere in the government! The truth is, getting permission to use the national forest site, arranging air and ground transportation, finding $50,000 to pay the air freight, and attending to myriad other physical and monetary obstacles, took us the better part of nine months.

Jacobs Conclusion 6: "What we photographed that day was the first terrestrial demonstration of what has come to be called S.D.I. or 'Star Wars.' The demonstration was put on for our benefit for some reason by extraterrestrials." Then what was the reason, and why did nothing come of it? No, the *terrestrial* demonstration period was so fruitful and successful that we established a permanent site at Anderson Peak above Big Sur!

FINDING THE "REAL" RV

What had we really photographed? Both the United States and the USSR had ongoing research programs in the 1960s for defense against ballistic missiles and to develop options to outwit possible defenses. Omitting the technical details, what had happened on Buzzing Bee was that two decoys were fired off by small rocket charges on schedule, but some of the decoy packing material also trailed along and could be seen optically and also by certain kinds of radar. A little cloud of debris around each decoy warhead clearly gave away the false status, almost as well as coloring the decoys bright red.

This, of course, led to more than a little consternation at SAC Headquarters and in higher military circles. Although correctable by redesign, the alarm in the minds of the strategic analysts was that the Soviets could defeat our ICBM decoys by using a few telescopes on mountain peaks in the USSR and relaying information on which objects were decoys to the Soviet ICBM defense command center. An immediate concern was that, although few understood its significance, a raft of people at Vandenberg AFB had seen the data. Vulnerability of a major weapons system is normally classified Top Secret. How could this matter be kept from leaking out?

ISSUE RESOLVED

As might be expected, the military reaction came swiftly. Everyone who was at the telescope site or had seen the film had to be identified. All, including Jacobs and myself, had to be questioned on what they had seen and what they thought it meant. Each was cautioned not to mention what was on the film to anyone and not to discuss it with others—even fellow workers who had originally seen it at the same time! None of us had more than a guess at the meaning, and the civilian intelligence experts who did the "debriefing" gave no hints.

Weeks later, my clearance level was increased to allow me to see the films again and analyze them. I don't think Bob Jacobs ever gained the required clearance. The people later assigned to operate the equipment and carry the films around were subsequently cleared to the required level. The Top Secret film was marked for downgrading and declassification after twelve years, but its utility was over after a few months. Top Secret storage is too difficult and expensive for keeping items of dubious worth, and the film and related materials were all destroyed long before the twelve years were up. Only a few of us even remember the incident today, and Bob Jacobs is being both safe and cagey in observing that the Air Force denies the existence of the film or other hard evidence.

The photo site established on Anderson Peak has undergone many changes and improvements over the years, and has continued to collect data during ICBM launches of high value to national defense. Much of the photography has needed security protection and the processes are in place to provide it without fanfare. There has never been a repetition of the security panic that followed the events of September 22, 1964, when Buzzing Bee literally and figuratively lit up the sky over the Pacific.

REFERENCES

George, Kingston A. 1964. *Preliminary Report on Image Orthicon Photography from Big Sur.* Headquarters 1st Strategic Aerospace Division Operations Analysis Staff Study, October 13.

Jacobs, Bob. 1989. "Deliberate Deception: The Big Sur UFO Filming." *MUFON UFO Journal,* no. 249 (January).

25. UFO "DOGFIGHT": A BALLOONING TALE

Joe Nickell

"UFO Fires on Louisville, Ky. Police Chopper" was the headline on the *Weekly World News*'s May 4 cover story, complete with fanciful illustration. But if the tabloid account seemed overly sensational in describing the "harrowing two-minute dogfight"—before vanishing into the night—it was only following the lead of the respected *Louisville Courier-Journal.* The *Courier* had used similar wording in relating the February 26 incident (which had not been immediately made public), headlining its front-page story of March 4: "UFO Puts on Show: Jefferson [County] Police Officers Describe Close Encounter."

Unfortunately, the *Weekly World News* did not cite the Courier's follow-up report that explained the phenomenon. Yet the tabloid's tale contained numerous clues that might have tipped off an astute reader. The first sighting was of what looked like "a fire" off to the patrol craft's left; the "pear-shaped" UFO was seen in the police spotlight "drifting back and forth like a balloon on a string"; after circling the helicopter several times, the object darted away before zooming back to shoot the "fireballs" (which fortunately "fizzled out before they hit"); and then—as the helicopter pilot pushed his speed to over 100 mph—the UFO "shot past the chopper, instantly climbing hundreds of feet," only to momentarily descend again before flying into the distance and disappearing. That the "flowing" object was only "about the size of a basketball" and that it had "hovered" before initially approaching the helicopter were additional clues from the original *Courier* account that the tabloid omitted.

This article originally appeared in the *Skeptical Inquirer* 18 (Fall 1993). Reprinted with permission.

The *Courier*'s follow-up story of March 6 was headed "A Trial Balloon?" It pictured Scott Heacock and his wife, Conchys, demonstrating how they had launched a hot-air balloon Scott had made from a plastic dry-cleaning bag, strips of balsa wood, and a dozen birthday candles—a device familiar to anyone who has read Philip J. Klass's *UFOs Explained* (Vintage Books, 1976, pp. 28–34, plates 2a and 2b). No sooner had the balloon cleared the trees, said Heacock, than the county police helicopter encountered it and began circling, shining its spotlight on the glowing toy.

The encounter was a comedy of errors and misperceptions. Likened to a cat chasing its tail, the helicopter was actually pushing the lightweight device around with its prop wash, In fact, as indicated by the officers' own account, the UFO zoomed away in response to the helicopter's sudden propulsion—behavior consistent with a lightweight object. As to the "fireballs," they may have been melting, flaming globs of plastic, or candles that became dislodged and fell, or some other effect. (Heacock says he used the novelty "relighting" type of birthday candies as a safeguard against the wind snuffing them out. Such candies may sputter, then abruptly reflame.)

Although one of the officers insisted the object he saw that night traveled at speeds too fast for a balloon, he seems not to have considered the effects of the helicopter's prop-wash propulsion. Contacted by psychologist and skeptical investigator Robert A. Baker, the other officer declined to comment further, except to state his feeling that the whole affair had been "blown out of proportion" by the media. Be that as it may, a television reporter asked Scott Heacock how certain he was that his balloon was the reported UFO. Since he had witnessed the encounter and kept the balloon in sight until it was caught in the police spotlight, he replied: "I'd bet my life on it." To another reporter, his Mexican wife explained: "I'm the only alien around here."

26. THAT'S ENTERTAINMENT! TV'S UFO COVERUP

Philip J. Klass

Don't be surprised or shocked if you discover that a good friend—a well-educated, intelligent person—believes in UFOs, or that he or she suspects that the U.S. government recovered a crashed extraterrestrial craft and ET bodies in New Mexico and has kept them under wraps for nearly half a century. Don't be surprised if your respected friend, or a member of your own family, is convinced that ETs are abducting thousands of Americans and subjecting them to dreadful indignities.

The really surprising thing is that *you* do not believe in crashed saucers, alien abductions, and government coverup if you spend even a few hours every week watching TV. There are many TV shows that promote belief in the reality of UFOs, government coverup, and alien abductions. And they attract very large audiences—typically tens of millions of viewers. Often they are broadcast a second, possibly even a third time.

TV has become the most pervasive means of influencing what people believe. That explains why companies spend billions of dollars every year on TV advertising to convince the public that Brand X beer tastes best, that you should eat Brand Y cereal, and that a Brand Z automobile is the world's best.

According to a recent survey reported in *Business Week* magazine, our children spend nearly twice as much time watching TV as they do in school.

Consider the problem that TV created for the Audi 5000 automobile and the claim that the car would suddenly accelerate and crash into the

---

This article originally appeared in the *Skeptical Inquirer* 20 (November/December 1996). Reprinted with permission.

front of an owner's garage when the automatic transmission was in neutral. The Audi 5000 was introduced in 1978, and during the next four years only thirteen owners complained of a mysterious sudden acceleration incident. Then, in November 1986, CBS featured the alleged Audi 5000 problem on its popular "60 Minutes" show. During the next month, some fourteen hundred people claimed that their Audi 5000s had experienced sudden acceleration problems (P. J. O'Rourke, *Parliament of Whores,* Atlantic Monthly Press, 1991, pp. 86–87). Subsequent investigation by the National Transportation Safety Board revealed that the problem was the result of driver error—stepping on the accelerator when they intended to step on the brake.

Here's another example: several years ago, a man who claimed he had found a hypodermic needle in a Pepsi-Cola can became an instant celebrity when he appeared on network TV news to describe his amazing discovery. Within several weeks, roughly fifty other persons around the country claimed they too had discovered hypodermic needles in Pepsi-Cola cans. Investigation showed all these reports were spurious.

TV's brainwashing of the public on UFOs occurs not only on NBC's "Unsolved Mysteries" and Fox network's "Sightings," but also on more respected programs such as CBS's "48 Hours" and ones hosted by CNN's Larry King.

Why pick on the TV networks? Cannot the same criticism be leveled at the print media? No. Generally, even cub reporters know that when writing an article on a controversial subject they should try to present both sides of the issue. If they fail to do so, their older and wiser managing editors will remind them. An article may devote 60 or 70 percent of its content to pro-UFO views, but with TV the pro-UFO content typically runs 95 percent—or higher.

TV news programs *do* try to offer viewers an even-handed treatment of controversial subjects. Thus it is not surprising that many viewers assume they are getting an equally balanced treatment in TV shows that follow the news, such as "Unsolved Mysteries" and "Sightings." This is especially true when the show is CBS's "48 Hours," hosted by news anchor Dan Rather.

This "schizophrenic" policy would be less troubling if such TV programs were required to carry a continuous disclaimer, such as "This program is providing you with a one-sided treatment of a controversial issue. It is intended solely to entertain you," or at least if such a disclaimer were voiced by the host at the beginning and the end of such a program. But alas, at best there is only a brief disclaimer which typically says: "The following is a controversial subject."

Consider a typical NBC "Unsolved Mysteries" show dealing with with the Roswell "crashed-saucer" incident. This show, which aired September 18, 1994, included an appearance by me. Prior to the taping of my interview, I gave the producer photocopies of once top-secret and secret air force documents that had never before been seen on TV and that provided important new evidence that a flying saucer had not crashed in New Mexico.

These documents, dating back to late 1948, revealed that if an ET craft was recovered from New Mexico in July 1947, nobody informed top Pentagon intelligence officials who should have been the first to know. One of these top-secret documents, dated December 10, 1948, more than a year and a half after the alleged recovery of an ET craft and "alien" bodies, showed that top air force and navy intelligence officials then believed that UFOs might be Soviet spy vehicles.

When the hour-long "Unsolved Mysteries" show aired, I appeared for only twenty seconds to discuss the early history of the UFO era. Not one of the once top-secret and secret documents, which disproved the Roswell myth, or my taped references to these documents, was used.

On October 1, 1994, the famed Larry King aired a two-hour special program on the TNT cable network. It's title was "UFO Coverup? Live From Area 5l." (Area 51 is part of an air force base in Nevada where new aircraft and weapons are tested. UFO believers allege that one can see alien spacecraft flying over the area and that the government has secret dealings and encounters with aliens there.)

Approximately one hour—half the two-hour program—was broadcast live from Nevada. For this hour, four pro-UFO guests were allowed to make wild claims, without a single live skeptic to respond. To give viewers the illusion of "balance," the show included pretaped interviews with Carl Sagan and with me. Sagan appeared in five very brief segments, averaging less than fifteen seconds each, for a total of one and one-quarter minutes. I appeared in four brief segments for a total airtime of one and a half minutes.

So during the two-hour show, the audience was exposed to less than three minutes of skeptical views on UFOs, crashed saucers, and government coverup. And because Sagan and I were taped many weeks earlier, neither of us could respond to nonsense spouted by the four UFO promoters who appeared live for an hour.

Some weeks earlier, when I went to the studio for my taped interview for this Larry King show, I handed producer Tom Farmer photocopies of the same once top-secret and secret documents I had given to "Unsolved Mysteries." Once again I stressed that these documents had never before appeared on any television show. Yet not one of these documents was shown during the two-hour program.

Near the end of the program, Larry King summed up the situation in the following words: "Crashed saucers. Who knows? But clearly the government is withholding something. . . ." In fact, it was Larry King and his producer who were withholding the hard data that would show that the government is not involved in a crashed-saucer coverup.

Larry King ended the program with these words: "We hope that you learned a lot tonight and that you found it both entertaining and informative at the same time."

If you were looking for a truly "informative" program on UFOs, you'd expect to find it on the "Science Frontiers" program broadcast on The Learning Channel, right? Wrong!

Last spring, The Learning Channel's "Science Frontiers" program aired a one-hour program titled "UFO." Not one of the many "UFO experts" interviewed on the program was a skeptic. The British producer sent a film crew to Washington—where I live—to interview pro-UFOlogist Fred Whiting, who was given nearly three minutes of airtime. Whiting assured the viewers: "There is indeed a coverup." But I was *not* invited to be interviewed.

In early 1994, I received a phone call from a producer of the CBS show "48 Hours," saying they were producing a segment on the Roswell crashed saucer and would like to come down from New York in mid-April to interview me.

In late March 1994, I visited Roswell in connection with a new crashed-saucer book that was making its debut there. Not surprisingly, the CBS film crew from "48 Hours" was on hand and they did a brief interview with me. In an effort to inform the viewers of the 1948 top-secret document, I pulled it out of my pocket and held it up in front of the CBS camera. And I promised to provide the producer with more such documents, never before shown on TV, when they came to Washington for the more lengthy interview.

CBS never came to Washington for my interview. And when the show later aired, with Dan Rather as its host, CBS opted *not* to include any of the brief interview with me in Roswell—holding up the once top-secret document.

Young children, and their parents, will experience similar "brainwashing" when they visit Disney World's new "Tomorrowland" in Orlando. A new dynamic exhibit is called "Alien Encounters and Extra-TERRORestrial Experience." To encourage parents and children to visit the new UFO exhibit, in mid-March 1995 Walt Disney Inc. broadcast a one-hour TV show on ABC titled "Alien Encounters from New Tomorrowland."

The show began by showing several brief home-video segments of bogus "UFOs" while the narrator intoned: "This is not swamp gas. It is not a flock of birds. This is an actual spacecraft from another world, piloted by alien intelligence. . . . Intelligent life from distant galaxies is now attempting to make open contact with the human race. Tonight we will show you the evidence."

The Disney show included the Roswell crashed-saucer case with considerable emphasis on government coverup. At one point, the narrator noted that Jimmy Carter had had a UFO sighting prior to becoming president. The narrator added: "Later, when he assumed the office of president . . . his staff attempted to explore the availability of official investigations into alien contacts."

Then, as the camera rapidly panned a typewritten document, it zoomed in on the words "no jurisdiction," and the narrator said: "As this internal government memo illustrates, there are some security secrets outside the jurisdiction even of the White House." The implication was that even the president did not have access to UFO secrets.

In reality, the memo was an FBI response to a White House inquiry about FBI involvement in investigating UFOs. The memo said that the FBI had "no jurisdiction" to investigate UFO reports and referred the White House to the air force. But the camera panned and zoomed so fast no viewer could read the memo.

Near the end of the program the narrator said: "Statistics indicate a greater probability that you will experience extraterrestrial contact in the next five years than the chances you will win a state lottery. But how do you prepare for such an extraordinary event? At Tomorrowland in Disney World, scientists and Disney engineers have brought to life a possible scenario that helps acclimate the public to their inevitable alien encounters."

More recently, Walt Disney Inc. has purchased the ABC television network. I won't be surprised if Disney and ABC use UFOs to attract more viewers.

For the tiny handful of those who produce TV and radio shows dealing with claims of the paranormal who truly want to provide their audience with both sides, the Committee for the Scientific Investigation of Claims of the Paranormal (CSICOP) is an invaluable resource in providing the names of experienced skeptics. The same is true for print-media reporters. If TV shows on UFOs are 95 percent "loaded" to promote belief, without CSICOP they would be 100 percent loaded.

Part Five
ALIEN ABDUCTIONS

27. 3.7 MILLION AMERICANS KIDNAPPED BY ALIENS?

Lloyd Stires

I recently received a sixty-page booklet, *Unusual Personal Experiences: An Analysis of the Data from Three National Surveys,* by Budd Hopkins, an artist and the author of *Intruders*; David Jacobs, associate professor of history at Temple University; and Ron Westrum, professor of sociology at Eastern Michigan University. It reports the results of a privately funded nationwide survey conducted for the authors by the Roper Organization, the purpose of which was to estimate the number of Americans who have been abducted by aliens. The authors' introduction states that this report is being sent to mental-health professionals in the hope that it will lead to more humane treatment of people suffering from "UFO abduction syndrome." According to the authors, psychologists at present treat people who believe they have been kidnapped by aliens as if they were mentally ill. They suggest that therapists should believe abductees and treat them as they would people with posttraumatic stress disorder, such as combat veterans and victims of family violence. (In fact, they draw a questionable analogy between UFO abductees and victims of child abuse, who also were not always believed by mental-health professionals.)

A random sample of 5,947 American adults participated in the survey. The sampling and data-collection methodology appears adequate, comparable to other national surveys. The difficulty lies with the questions and the assumptions underlying their interpretation. The authors claim that you can't ask people directly whether they have been abducted by aliens: first, because some of the victims have repressed the experience

This article originally appeared in the *Skeptical Inquirer* 17 (Winter 1993). Reprinted with permission.

and, second, because some of those who remember their abductions have been ridiculed for talking about them and are reluctant to discuss them with strangers. Therefore, they tried to measure abduction indirectly, using five questions about specific events commonly reported by abductees. (Why should people who are repressing or concealing their abductions nonetheless respond to these five questions? Presumably, they are less threatening. If this assumption is false, then by the authors' logic the number of alien abductees will be underestimated.)

The five questions—all preceded by "How often has this occurrence happened to you?"—are as follows. (The percentages in brackets represent those who said this had happened to them at least once.)

> Waking up paralyzed with a sense of a strange person or presence or something else in the room. [18%]
>
> Experiencing a period of time of an hour or more in which you were apparently lost, but you could not remember why or where you had been. [13%]
>
> Feeling that you were actually flying through the air although you didn't know how or why. [10%]
>
> Seeing unusual lights or balls of light in a room without knowing what was causing them or where they came from. [8%]
>
> Finding puzzling scars on your body and neither you nor anyone else remembering how you received them or where you got them. [8%]

These questions were selected from interviews with people who believe they have been abducted by aliens. Many of them report similar scenarios. They wake up immobilized in a room surrounded by alien creatures ("Small, gray-skinned, hairless figures" with large eyes) and balls of light. They are levitated to a metallic spacecraft, where they are stripped and subjected to medical examinations (aliens take an unusual interest in the genitals of abductees) that sometimes leave scars. Afterward, they are unable to account for their lost time.

Respondents who reported having four out of these five experiences were considered probable abductees. However, the authors recognized that they might simply be measuring suggestibility, so they added another part to the question as a control:

> Hearing or seeing the word *trondant* and knowing that it has a secret meaning for you.

Only 1 percent said they recognized this nonexistent word, and these people were discounted as probable abductees. After these false positives were eliminated 2 percent of the sample (119 people) met the criteria as probable abductees. Since the respondents were randomly sampled from the total population of 185 million American adults, the authors infer that 3.7 million Americans have probably been abducted by aliens.

No evidence is presented for the validity of these five questions. That is, we have no assurance that they measure what they are supposed to measure. The authors assume that alien abductees are likely to answer yes to four of the five questions, while nonabductees are not. Obviously, there is no group of known abductees to whom the questions have been posed. In fact, the authors made no attempt to validate the much weaker assumption that people *who believe they have been abducted by aliens* are more likely than other people to agree to these items. Therefore, the questions are useless for the stated purpose. (They could have asked respondents directly whether they believed they had been abducted by aliens *at the end of the survey,* where it would not have contaminated the responses to the other questions.)

How can we explain the high percentages of people who reported having these unusual experiences and the 2 percent who reported having four out of five of them? Several possibilities exist.

1. Maybe the percentages are not surprising, considering the number of people who hold paranormal beliefs (Gallup and Newport 1991). Note that it is theoretically possible to answer yes to all five questions without ever entertaining a UFO-abduction scenario.

2. The authors may still be measuring suggestibility. The control question may be ineffective since familiarity with the word trondant is not as interesting or appealing as the other, paranormal beliefs.

3. Respondents may have been confused about the meaning of some of the questions. For example, those responding yes to the item about flying through the air may have been reporting dreams about flying, which are fairly common. The authors claim that the word *actually* in the question precludes this interpretation and that the phrase "although you didn't know how or why" eliminates the reporting of airplane rides, falls, and so on. However, this assumes that subjects are very attentive to the questions and conscientious in their responses. There are similar problems with the other four items.

The authors seem impressed with the fact that so many victims of UFO-abduction syndrome report similar experiences. However, the existence of a standard alien-abduction scenario can be explained by their common exposure to books, films, and television dramas with this plot.

I wonder whether the motives of the authors and publishers of this report are completely altruistic. The booklet contains a reply card on which you can indicate your interest in future conferences and workshops on the subject.

REFERENCES

Gallup, G. H., and F. Newport. 1991. "Belief in Paranormal Phenomena among Adult Americans." *Skeptical Inquirer* 15: 137–46.

Hopkins, B., D. M. Jacobs, and R. Westrum. 1992. *Unusual Personal Experiences: An Analysis of the Data from Three National Surveys.* Bigelow Holding Corporation, 4640 South Eastern, Las Vegas, NV 89119.

28. ADDITIONAL COMMENTS ABOUT THE "UNUSUAL PERSONAL EXPERIENCES SURVEY"

Philip J. Klass

The Roper survey was conducted to try to determine how many American adults may have experienced "UFO abductions." But the 11 questions asked were framed by Budd Hopkins and David Jacobs—the chief promoters of the claim that ETs are abducting Earthlings.

Only eighteen out of the 5,947 persons surveyed (0.3%) reported *all five* of the "key indicator" experiences—which would mean that "only" 560,000 American adults had experienced UFO abduction. So Hopkins and Jacobs decided that if anyone answered yes to four out of the five experiences, this qualified him or her as a "probable abductee." When Hopkins and Jacobs used this relaxed criterion, the Roper survey showed that 2 percent of those surveyed qualified as "probable abductees," which corresponds to 3.7 million "probable abductees"—a much more impressive figure.

If their interpretation of the Roper data were correct, consider the implications. If one assumes that UFO abductions began in the fall of 1961 with Betty and Barney Hill, and since then ETs have abducted 3.7 million Americans, this means that an average of nearly 340 Americans have been abducted every day during the past thirty years. Because most UFO abductions (allegedly) occur at night, this means that (on average) every two minutes during every night of the past thirty years an American has been abducted. . . .

It is regrettable (but not surprising) that Hopkins and Jacobs did not

---

This article originally appeared in the *Skeptical Inquirer* 17 (Winter 1993). Reprinted with permission. These remarks were excerpted by permission from Philip J. Klass's *Skeptics UFO Newsletter* 16 (July 1992).

include any survey questions asking how many books dealing with UFOs the subject had read or how many television shows dealing with UFOs and UFO-abductions had been seen to assess their possible influence. . . .

In discussing the results of the survey, Hopkins, Jacobs, and Ron Westrum (a sociology professor at Eastern Michigan University) gloss over the fact that 11 percent of those surveyed say they've seen ghosts and 14 percent (26 million persons) report "feeling as if you left your body."

The survey indicates that 11 percent of those surveyed (corresponding to more than 20 million persons) said that they had seen a ghost, and 3 percent (5.5 million persons) said they had seen a ghost more than twice. But only 7 percent (13 million persons) reported having had UFO sightings and 1 percent (1.9 million) reported more than two sightings.

This might seem to show that the United States is being visited by more ghosts than UFOs. But a 1990 telephone survey of 1,236 American adults, conducted by the Gallup organization ("Belief in Paranormal Phenomena Among Adult Americans," by George H. Gallup and Frank Newport, *Skeptical Inquirer,* Winter 1991), showed that 14 percent of those polled had seen a UFO, while only 9 percent reported seeing a ghost. Gallup's 14 percent UFO-sighting figure is twice Roper's 7 percent. . . .

Hopkins and Jacobs are surprised that the highest number of yes responses to the five "key indicator" questions was the 18 percent for the one which asked about "waking up paralyzed and sensing the presence of a strange figure." They acknowledge the occurrence of "hypnogogic hallucinations" by perfectly normal persons when falling asleep, or "hypnopompic hallucinations" when awakening, in which a person reports feeling paralyzed. But by adding the provisions of sensing "a strange person or presence or something else in the room," Hopkins and Jacobs claim this excludes a possible hypnopompic/hypnogogic explanation.

If Hopkins or Jacobs had read the Summer 1988 *Skeptical Inquirer,* they would know that their claim is false. The Winter 1987–88 *Skeptical Inquirer* carried an article on hypnopompic/hypnogogic hallucinations, authored by Robert A. Baker, a seasoned professor of psychology at the University of Kentucky. Baker's article prompted a number of readers to write, describing their own hypnopompic/hypnogogic experiences, some of which were published in the Letters section of the Summer 1988 issue of *Skeptical Inquirer*. . . .

Experiments conducted by psychologists Sheryl C. Wilson and T. X. Barber indicate that an estimated 4 percent of adult Americans are "fantasy-prone individuals." Such persons "fantasize a large part of the time" and "typically 'see' . . . and fully experience what they fantasize," according to Wilson and Barber. Results of the Roper survey suggest their 4 percent figure may be low. . . .

Many psychotherapists will be impressed by the fact that the introduction to the Roper survey report was written by John E. Mack, sixty-two, professor of psychiatry at Harvard Medical School and former head of its Psychiatric Department. In Mack's introduction, he did not mention that he recently signed a $200,000 contract with Scribner's to write a book on UFO abductions. . . .

Funds to conduct the Roper survey, publish the sixty-four-page report on the results, and mail it to nearly 100,000 psychiatrists, psychologists, and other mental-health professionals shortly before the CBS-TV miniseries "Intruders" dealing with UFO abductions was broadcast May 17–19, were supplied by Robert Bigelow, a wealthy Las Vegas businessman, and an "anonymous donor" (whose name is Hans-Adam von Lichtenstein, from the country whose name he bears).

We suggest that Bigelow and von Lichtenstein fund a similar survey in a country in which UFOs and UFO abductions have *not* received such wide promotion on television—for example, Bulgaria. We predict that a far smaller percentage of Bulgarians will qualify as "probable abductees."

29.
THE ALIENS AMONG US: HYPNOTIC REGRESSION REVISITED

Robert A. Baker

For the average person walking down the aisle of a modern bookstore or passing through the checkout lane at the nearest supermarket, it would be easy to conclude that aliens from outer space not only are here but also have joined the Baptist church, have put their kids in school, and belong to the Rotary Club. This conclusion is demanded by the recent rash of nonfiction books about UFO contacts, encounters of the third kind, and human abductions by little gray men from outer space or some other parallel universe. Typical of these tomes are *Communion,* by Whitley Strieber; *Intruders,* by Budd Hopkins; and *Light Years: An Investigation into the Extraterrestrial Experience of Eduard Meier,* by Gary Kinder. (See reviews of the Strieber and Hopkins books in the Fall 1987 *Skeptical Inquirer.*) According to these and other UFO pundits, abductions by "little gray aliens" are so prevalent they will soon become commonplace and generally accepted as a fact of life by a now skeptical public and press.

My friends and colleagues and I, however, are beginning to believe that we have Alien B.O. or something worse, because none of us has been contacted, interviewed, briefed, threatened, kidnapped, or physically examined by any of the little folk. We, sadly enough, have not even had our car stalled by one of their spaceships. It stalls on occasion, but the problem lies in Detroit rather than with the aliens. Could all this alien activity going on around us be overlooked by responsible authorities?

To impress the general reader, all three authors have taken great pains

---

This article originally appeared in the *Skeptical Inquirer* 12 (Winter 1987–88). Reprinted with permission.

to give as much credibility and authenticity as possible to their claims. Strieber not only took a lie-detector test but also had a psychiatrist write a statement attesting to his sanity.[1] Kinder had professional photographers examine a number of Meier's photographs and also had an IBM metallurgist endorse the unusual quality of a metal fragment from the purported spaceship. To Kinder's credit, however, he admits that he is skeptical about some of Meier's claims—particularly that of journeying back in time and talking to Jesus Christ. As for Hopkins, he not only consulted a number of psychologists and psychiatrists (he even found an abductee among them) but also had medical specialists corroborate the correctness of the medical techniques the aliens used to examine their human subjects. Just why aliens should copy human medical approaches is an unanswered question.

One would have thought that Philip Klass's (1981) devastating attack on abductee claims, published in this journal, coupled with Robert Sheaffer's (1981) brilliant and calmly reasoned work, *The UFO Verdict,* and Douglas Curran's (1985) *In Advance of the Landing: Folk Concepts of Outer Space,* along with William R. Corliss's (1983) *Handbook of Unusual Natural Phenomena,* would have given the true believers pause and would have dampened somewhat their extravagant claims. But, like a rubber ball, they keep bouncing back.

Sheaffer and Corliss offer credible and scientific explanations of 99 percent or more of the strange lights in the sky, whereas Curran's extensive catalog of aberrant human believers suggests that the true aliens in our midst are not from outer space or a parallel dimension but are our fellow *Homo sapiens* from the edge of town. If you wish to see some excellent photos of aliens, study the pictures and read the biographies in Curran's book.[2] Truly, the aliens and the alienated are already among us and have been for a long while, differing from the majority of other Americans only in the extreme nature of their beliefs and convictions. Klass's continuing excellent work on UFO demystification highlights the significance of hypnotic regression in the abductee belief system. For hypnotic regression and the personality pattern Wilson and Barber (1983) call "fantasy-prone," as well as the behavior of individuals undergoing hypnogogic and hypnopompic experiences, furnish, we believe, complete and credible explanations to most—if not all—accounts of UFO contacts and abductions past and present.

Most people seem unaware of the fact that there is an already well established branch of psychology, anomalistic psychology, that deals specifically with the kind of experiences had by Strieber, Meier, and the other UFO abductees. This psychology provides naturalistic and satisfying explanations for the entire range of such behaviors. Let us examine these explanations a little more closely and in a little more detail.

HYPNOSIS AND HYPNOTIC REGRESSION

In France in the 1770s, when Mesmerism was in its heyday, the king appointed two commissions to investigate Mesmer's activities. The commissions included such eminent men as Benjamin Franklin, Lavoisier, and Jean-Sylvain Bailly, the French astronomer. After months of study the report of the commissioners concluded that it was imagination, not magnetism, that accounted for the swooning, trancelike rigidity of Mesmer's subjects. Surprisingly enough, this conclusion is still closer to the truth about hypnosis than most of the modern definitions found in today's textbooks.

So-called authorities still disagree about "hypnosis." But whether it is or is not a "state," there is common and widespread agreement among all the major disputants that "hypnosis" is a situation in which people set aside critical judgment (without abandoning it entirely) and engage in make-believe and fantasy; that is, they use their imagination (Sarbin and Andersen 1967; Barber 1969; Gill and Brenman 1959; Hilgard 1977). As stated earlier, there are great individual differences in the ability to fantasize, and in recent years many authorities have made it a *requirement* for any successful "hypnotic" performance. Josephine Hilgard (1979) refers to hypnosis as "imaginative involvement," Sarbin and Coe (1972) term it "believed-in imaginings," Spanos and Barber (1974) call it "involvement in suggestion-related imaginings," and Sutcliffe (1961) has gone so far as to characterize the hypnotizable individual as someone who is "deluded in a descriptive, nonpejorative sense" and he sees the hypnotic situation as an arena in which people who are skilled at make-believe and fantasy are provided with the opportunity and the means to do what they enjoy doing and what they are able to do especially well. Even more recently Perry, Laurence, Nadon, and Labelle (1986) concluded that "abilities such as imagery/imagination, absorption, disassociation, and selective attention underlie high hypnotic responsivity in yet undetermined combinations." The same authors, in another context dealing with past-lives regression, also concluded that "it should be expected that any material provided in age regression (which is at the basis of reports of reincarnation) may be fact or fantasy, and it is most likely an admixture of both." The authors further report that such regression material is colored by issues of confabulation, memory creation, inadvertent cueing, and the regressee's current psychological needs. (See also Nicholas Spanos's article in the *Skeptical Inquirer* 12, Winter 1987–88.)

CONFABULATION

Because of its universality, it is quite surprising that the phenomenon of confabulation is not better known. Confabulation, or the tendency of ordinary, sane individuals to confuse fact with fiction and to report fantasized events as actual occurrences, has surfaced in just about every situation in which a person has attempted to remember very specific details from the past. A classical and amusing example occurs in the movie *Gigi,* in the scene where Maurice Chevalier and Hermione Gingold compare memories of their courtship in the song "I Remember It Well." We remember things not the way they really were but the way we would have liked them to have been.

The work of Elizabeth Lotus and others over the past decade has demonstrated that the human memory works not like a tape recorder but more like the village storyteller—i.e., it is both creative and recreative. We can and we do easily forget. We blur, shape, erase, and change details of the events in our past. Many people walk around daily with heads full of "fake memories." Moreover, the unreliability of eyewitness testimony is not only legendary but well documented. When all of this is further complicated and compounded by the impact of suggestions provided by the hypnotist plus the social-demand characteristics of the typical hypnotic situation, little wonder that the resulting recall on the part of the regressee bears no resemblance to the truth. *In fact, the regressee often does not know what the truth is.*

Confabulation shows up without fail in nearly every context in which hypnosis is employed, including the forensic area. Thus it is not surprising that most states have no legal precedents on the use of hypnotic testimony. Furthermore, many state courts have begun to limit testimony from hypnotized witnesses or to follow the guidelines laid down by the American Medical Association in 1985 to assure that witnesses' memories are not contaminated by the hypnosis itself. For not only do we translate beliefs into memories when we are wide awake, but in the case of hypnotized witnesses with few specific memories the hypnotist may unwittingly suggest memories and create a witness with a number of crucial and vivid recollections of events that never happened, i.e., pseudo-memories. It may turn out that the recent Supreme Court decision allowing the individual states limited use of hypnotically aided testimony may not be in the best interests of those who seek the truth. Even in their decision the judges recognized that hypnosis may often produce incorrect recollections and unreliable testimony.

There have also been a number of clinical and experimental demonstrations of the creation of pseudo-memories that have subsequently come to be believed as veridical. Hilgard (1981) implanted a false memory of an experience connected with a bank robbery that never occurred. His subject found the experience so vivid that he was able to select from a series of photographs a picture of the man he thought had committed the robbery. At another time, Hilgard deliberately assigned two concurrent—though spatially different—life experiences to the same person and regressed him at separate times to *that date.* The individual subsequently gave very accurate accounts of both experiences, so that anyone believing in reincarnation who reviewed the two accounts would conclude the man *really had* lived the two assigned lives.

In a number of other experiments designed to measure eyewitness reliability, Lotus (1979) found that details supplied by others invariably contaminated the memory of the eyewitness. People's hair changed color, stop signs became yield signs, yellow convertibles turned to red sedans, the left side of the street became the right-hand side, and so on. The results of these studies led her to conclude, "It may well be that the legal notion of an independent recollection is a psychological impossibility." As for hypnosis, she says: "There's no way even the most sophisticated hypnotist can tell the difference between a memory that is real and one that's created. If a person is hypnotized and highly suggestible and false information is implanted in his mind, it may get embedded even more strongly. One psychologist tried to use a polygraph to distinguish between real and phony memory, but it didn't work. Once someone has constructed a memory, he comes to believe it himself."

CUEING: INADVERTENT AND ADVERTENT

Without a doubt, inadvertent cueing also plays a major role in UFO-abduction fantasies. The hypnotist unintentionally gives away to the person being regressed exactly what response is wanted. This was most clearly shown in an experimental study of hypnotic age regression by R. M. True in 1949. He found that 92 percent of his subjects, regressed to the day of their tenth birthday, could accurately recall the day of the week on which it fell. He also found the same thing for 84 percent of his subjects for their fourth birthday. Other investigators, however, were unable to duplicate True's findings. When True was questioned by Martin Orne about his experiment, he discovered that the editors of *Science,* where his report had appeared, altered his procedure section without his prior consent. True, Orne discovered, had inadvertently cued his subjects by following the unusual technique of asking them, "Is it Monday? Is it Tuesday? Is it Wednesday?" etc., and he monitored their responses by using a perpetual desk calendar in full view of all his subjects. Further evidence of the prevalence and importance of such cueing came from a study by O'Connell, Shor, and Orne (1970). They found that in an existing group of four-year-olds not a single one knew what day of the week it was. The reincarnation literature is also replete with examples of such inadvertent cueing. Ian Wilson (1981), for example, has shown that hypnotically elicited reports of being reincarnated vary as a direct function of the hypnotist's belief about reincarnation. Finally, Laurence, Nadon, Nogrady, and Perry (1986) have shown that pseudo-memories were elicited also by inadvertent cueing in the use of hypnosis by the police.

As for advertent, or *deliberate,* cueing, one of my own studies offers a clear example. Sixty undergraduates divided into three groups of twenty each were hypnotized and age-regressed to previous lifetimes. Before each hypnosis session, however, suggestions very favorable to and supportive of past-life and reincarnation beliefs were given to one group; neutral and noncommittal statements about past lives were given to the second group; and skeptical and derogatory statements about past lives were given to the third group. The results clearly showed the effects of these cues and suggestions. Subjects in the first group showed the most past-life regressions and the most past-life productions; subjects in the third group showed the least (Baker 1982).

Regression subjects take cues as to how they are to respond from the person doing the regressions and asking the questions. If the hypnotist is a believer in UFO abductions the odds are heavily in favor of him eliciting UFO-abductee stories from his volunteers.

FANTASY-PRONE PERSONALITIES AND PSYCHOLOGICAL NEEDS

"Assuming that all you have said thus far *is* true," the skeptical observer might ask, "why would hundreds of ordinary, mild-mannered, unassuming citizens suddenly go off the deep end and turn up with cases of amnesia and then, when under hypnosis, all report nearly identical experiences?" First, the abductees are not as numerous as we are led to believe; and, second, even though Strieber and Hopkins go to great lengths to emphasize the diversity of the people who report these events, they are much more alike than these taxonomists declare. In an afterword to Hopkins's *Missing Time,* a psychologist named Aphrodite Clamar raises exactly this question and then adds, "All of these people seem quite ordinary in the psychological sense—*although they have not been subjected to the kind of psychological testing that might provide a deeper under-*

standing of their personalities" (emphasis added). And herein lies the problem. If these abductees were given this sort of intensive diagnostic testing it is highly likely that many similarities would emerge—particularly an unusual personality pattern that Wilson and Barber (1983) have categorized as "fantasy-prone." In an important but much neglected article, they report in some detail their discovery of a group of excellent hypnotic subjects with unusual fantasy abilities. In their words:

> Although this study provided a broader understanding of the kind of life experiences that may underlie the ability to be an excellent hypnotic subject, it has also led to a serendipitous finding that has wide implication for all of psychology—it has shown that there exists a small group of individuals (possibly 4% of the population) who fantasize a large part of the time, who typically "see," "hear," "smell," and "touch" and fully experience what they fantasize; and who can be labeled fantasy-prone personalities.

Wilson and Barber also stress that such individuals experience a reduction in orientation to time, place, and person that is characteristic of hypnosis or trance during their daily lives whenever they are deeply involved in a fantasy. They also have experiences during their daily ongoing lives that resemble the classical hypnotic phenomena. In other words, the behavior we would normally call "hypnotic" is exhibited by these fantasy-prone types (FPs) all the time. In Wilson and Barber's words: "When we give them 'hypnotic suggestions,' such as suggestions for visual and auditory hallucinations, negative hallucinations, age regression, limb rigidity, anesthesia, and sensory hallucinations, we are asking them to do for us the kind of thing they can do independently of us in their daily lives."

The reason we do not run into these types more often is that they have learned long ago to be highly secretive and private about their fantasy lives. Whenever the FPs do encounter a hypnosis situation it provides them with a social situation in which they are encouraged to do, and are rewarded for doing, what they usually do only in secrecy and in private. Wilson and Barber also emphasize that regression and the reliving of previous experiences is something that virtually all the FPs do naturally in their daily lives. When they recall the past, they relive it to a surprisingly vivid extent, and they all have vivid memories of their experiences extending back to their early years.

Fantasy-prone individuals also show up as mediums, psychics, and religious visionaries. They are also the ones who have many realistic "out of body" experiences and prototypic "near-death" experiences.

In spite of the fact that many such extreme types show FP characteristics, the overwhelming majority of FPs fall within the broad range of normal functioning. It is totally inappropriate to apply a psychiatric diagnosis to them. in Wilson and Barber's words: "It needs to be strongly emphasized that our subjects with a propensity for hallucinations are as well adjusted as our comparison group or the average person. It appears that the life experiences and skill developments that underlie the ability of hallucinatory fantasy are more or less independent of the kinds of life experience that leads to pathology." In general, FPs are "normal" people who function as well as others and who are as well adjusted, competent, and satisfied or dissatisfied as everyone else.

Anyone familiar with the the fantasy-prone personality who reads *Communion* will suffer an immediate shock of recognition. Strieber is a classic example of the genre: he is easily hypnotized; he is amnesiac; he has vivid memories of his early life, body immobility and rigidity, a very religious background, a very active fantasy life; he is a writer of occult and highly imaginative novels; he has unusually strong sensory experiences—particularly smells and sounds—and vivid dreams. More interesting still is the comment made by Strieber's wife during her questioning under hypnosis by Budd Hopkins (p. 197). In referring to some of Strieber's visions she says: "Whitley saw a lot of things that I didn't see at that time." "Did you look for it?" "Oh, no. Because I knew it wasn't real." "How did you know it wasn't real? Whitley's a fairly down-to-earth guy—" "No, he isn't." . . . "It didn't surprise you hearing Whitley, that he sees things like that [a bright crystal in the sky]?" "No." It seems if anyone really knows us well it's our wives. But even more remarkable are the correspondences between Strieber's alien encounters and the typical hypnopompic hallucinations to be discussed later.

It is perfectly clear, therefore, why most of the UFO abductees, when given cursory examinations by psychiatrists and psychologists, would turn out to be ordinary, normal citizens as sane as themselves. It is also evident why the elaborate fantasies woven in fine cloth from the now universally familiar UFO-abduction fable—a fable known to every man, woman, and child newspaper reader or moviegoer in the nation—would have so much in common, so much consistency in the telling. Any one of us, if asked to pretend that he had been kidnapped by aliens from outer space or another dimension, would make up a story that would vary little, either in its details or in the supposed motives of the abductors, from the stories told by any and all of the kidnap victims reported by Hopkins. As for the close encounters of the third kind and conversations with the little gray aliens described in *Communion* and *Intruders,* again, our imaginative tales would be remarkably similar in plot, dialogue, description, and characterization. The means of transportation would be saucer-

THE POWER OF SUGGESTION ON MEMORY

In my own work on hypnosis and memory, the power of suggestion on the evocation of false memories was clearly and dramatically evident. Sixty volunteers observed a complex visual display made up of photographs of a number of common objects, e.g., a television set, a clock, a typewriter, a book, and so on, and eight nonsense syllables. They were instructed to memorize the nonsense syllables in the center of the display and were given two minutes to accomplish it. Nothing was said about the common objects. Following a forty-minute delay the students were questioned about the nonsense syllables and the other objects on display. They were also asked to state their confidence in the accuracy of their answers. Some were questioned under hypnosis and others while they were wide awake.

As a secondary part of the study the extent of the student's suggestibility was also studied. This was done by asking them to report on the common objects (as well as their primary task of memorizing the nonsense syllables) and asking specific questions about objects that *were not on the display.* Since their attention was not directed at the objects *specifically,* they were of course unsure about what they saw and didn't see. Therefore, when they were asked the questions "What color was the sports car?" and "Where on the display was it located" they immediately assumed there must have been a sports car present or I wouldn't be asking the question. Similarly with a suggested lawnmower and calendar. Although 35 subjects reported the color of the suggested automobile in the hypnoidal condition, 34 reported the color while awake. Similarly, although 26 subjects reported the suggested lawnmower's color and position in the hypnoidal state, 27 reported its color and position while awake. For the nonexistent calendar, 24 reported the month and date while hypnotized, and 23 did so while awake.

As for suggestibility *per se* under all conditions, 50 out of 60 volunteers reported seeing something that wasn't there with a confidence level of 2 (a little unsure) or greater, while 45 out of 60 reported seeing something that wasn't there with a confidence level of 3 (sure) or greater, whereas 25 out of 60 reported seeing something that was not there with a confidence level of 4 (very sure) or greater. Finally, 8 out of the 60 reported something not there with a confidence level of 5 (absolute certainty). Interestingly enough, 5 of the 8 reported they were certain of the object's existence even though they were wide awake; and, when they were allowed to see the display again, they were shocked to discover their error (Baker, Haynes, and Patrick 1983).

shaped; the aliens would be small, humanoid, two-eyed, and gray, white, or green. The purpose of their visits would be: (1) to save our planet; (2) to find a better home for themselves; (3) to end nuclear war and the threat we pose to the peaceful life in the rest of galaxy; (4) to bring us knowledge and enlightenment; and (5) to increase their knowledge and understanding of other forms of intelligent life. In fact, the fantasy-prone abductees' stories would be much more credible if some of them, at least, reported the aliens as eight-foot-tall, red-striped octapeds riding bicycles and intent upon eating us for dessert.

Finally, what would or could motivate even the FPs to concoct such outlandish and absurd tales, tales that without fail draw much unwelcome attention and notoriety? What sort of psychological motives and needs would underlie such fabrications? Perhaps the best answer to this question is the one provided by the author-photographer Douglas Curran. Traveling from British Columbia down the West Coast and circumscribing the United States along a counterclockwise route, Curran spent more than two years questioning ordinary people about outer space. Curran writes:

> On my travels across the continent I never had to wait too long for someone to tell me about his or her UFO experience, whether I was chatting with a farmer in Kansas, Ruth Norman at the Unarius Foundation, or a cafe owner in Florida. What continually struck me in talking with these people was how positive and ultimately life-giving a force was their belief in outer space. Their belief reaffirmed the essential fact of human existence: the need for order and hope. It is this that establishes them—and me—in the continuity of human experience. It brought to me a greater understanding of Oscar Wilde's observation. "We are all lying in the gutter—but some of us are looking at the stars."

Jung (1969), in his study of flying saucers, first published in 1957, argues that the saucer represents an archetype of order, wholeness, deliverance, and salvation—a symbol manifested in other cultures as a sun wheel or magic circle. Further in his essay, Jung compares the spacemen aboard the flying saucers to the angelic messengers of earlier times who brought messages of hope and salvation—the theme emphasized in Strieber's *Communion.* Curran also observes that the spiritual message conveyed by the aliens is, recognizably, our own. None of the aliens Curran's contactees talked about advocated any moral or metaphysical belief that was not firmly rooted in the Judeo-Christian tradition. As Curran says, "Every single flying-saucer group I encountered in my travels incorporated Jesus Christ into the hierarchy of its belief system." No wonder Eduard Meier had to travel back in time and visit the Savior. Many theorists have long recognized that whenever world events prove to be psychologically destabilizing, men turn to religion as their only hope. Jung, again, in his 1957 essay, wrote: "In the threatening situation of the world today, when people are beginning to see that everything is at stake, the projection-creating fantasy soars beyond the realm of earthly organization and powers into the heavens, into interstellar space, where the rulers of human fate, the gods, once had their abode in the planets."

The beauty and power of Curran's portraits of hundreds of true UFO believers lies in his sympathetic understanding of their fears and frailties. As psychologists are well aware, our religions are not so much systems of objective truths about the universe as they are collections of subjective statements about humanity's hopes and fears. The true believers interviewed by Curran are all around us. Over the years I have encountered several. One particularly memorable and poignant case was that of a federal prisoner who said he could leave his body at will and sincerely believed it. Every weekend he would go home to visit his family while (physically) his body stayed behind in his cell. Then there was the female psychic from the planet Xenon who could turn electric lights on and off at will, especially traffic signals. Proof of her powers? If she drove up to a red light she would concentrate on it intently for thirty to forty seconds and then, invariably, it would turn green!

HYPNOGOGIC AND HYPNOPOMPIC HALLUCINATIONS

Another common yet little publicized and rarely discussed phenomenon is that of hypnogogic (when *falling asleep*) and hypnopompic (when *waking up*) hallucinations. These phenomena, often referred to as "waking dreams," find the individual suddenly awake, but paralyzed, unable to move, and most often encountering a "ghost." The typical report goes somewhat as follows, "I went to bed and went to sleep and then sometime near morning something woke me up. I opened my eyes and found myself wide awake but unable to move. There, standing at the foot of my bed was my mother, wearing her favorite dress—the one we buried her in. She stood there looking at me and smiling and then she said: 'Don't worry about me, Doris, I'm at peace at last. I just want you and the children to be happy.' " Well, what happened next? "Nothing, she slowly faded away." What did you do then? "Nothing, I just closed my eyes and went back to sleep."

There are always a number of characteristic clues that indicate a hypnogogic or hypnopompic hallucination. First, it always occurs before or after falling asleep. Second, one is paralyzed or has difficulty in moving; or, contrarily, one may float out of one's body and have an out-of-body experience. Third, the hallucination is unusually bizarre; i.e., one sees ghosts, aliens, monsters, and such. Fourth, after the hallucination is over the hallucinator typically goes back to sleep. And, fifth, the hallucinator is unalterably convinced of the "reality" of the entire experience.

In Strieber's *Communion* (pp. 172–75) is a classic, textbook description of a hypnopompic hallucination, complete with the awakening from a sound sleep, the strong sense of reality and of being awake, the paralysis (due to the fact that the body's neural circuits keep our muscles relaxed and help preserve our sleep), and the encounter with strange beings. Following the encounter, instead of jumping out of bed and going in search of the strangers he has seen, Strieber typically goes back to sleep. He even reports that the burglar alarm was still working—proof again that the intruders were mental rather than physical. Strieber also reports an occasion when he awakes and believes that the roof of his house is on fire and that the aliens are threatening his family. Yet his only response to this was to go peacefully back to sleep. Again, clear evidence of a hypnopompic dream. Strieber, of course, is convinced of the reality of these experiences. This too is expected. If he was not convinced of their reality, then the experience would not be hypnopompic or hallucinatory.

The point cannot be more strongly made that ordinary, perfectly sane and rational people have these hallucinatory experiences and that such individuals are in no way mentally disturbed or psychotic. But neither are such experiences to be taken as incontrovertible proof of some sort of objective or consensual reality. They may be subjectively real, but objectively they are nothing more than dreams or delusions. They are called "hallucinatory" because of their heightened subjective reality. Leaving no rational explanation unspurned, Strieber is nevertheless forthright enough to suggest at one point the possibility that his experiences indeed could be hypnopompic. Moreover, in a summary chapter he speculates, correctly, that the alien visitors could be "from within us" and/or "a side effect of a natural phenomenon . . . a certain hallucinatory wire in the mind causing many different people to have experiences so similar as to seem to be the result of encounters with the same physical phenomena" (p. 224).

Interestingly enough, these hypnopompic and hypnogogic hallucinations do show individual differences in content and character as well as a lot of similarity: ghosts, monsters, fairies, friends, lovers, neighbors, and even little gray men and golden-haired ladies from the Pleiades are frequently encountered. Do such hallucinations appear more frequently to highly imaginative and fantasy-prone people than to other personality types? There is some evidence that they do (McKellar 1957; Tart 1969; Reed 1972; Wilson and Barber 1983), and there can certainly be no doubt that Strieber is a highly imaginative personality type.

"MISSING" TIME?

As for the lacunae or so-called missing time experienced by all the UFO abductees, this too is a quite ordinary, common, and universal experience. Jerome Singer (1975) in his *Inner World of Daydreaming* comments:

> Are there ever any truly "blank periods" when we are awake? It certainly seems to be the case that under certain conditions of fatigue or great drowsiness or extreme concentration upon some physical act we may become aware that we cannot account for an interval of time and have no memory of what happened for seconds and sometimes minutes.

Graham Reed (1972) has also dealt with the "time-gap" experience at great length. Typically, motorists will report after a long drive that at some point in the journey they wake up to realize they have no awareness of a preceding period of time. With some justification, people still will describe this as a "gap in time," a "lost half-hour," or a "piece out of my life." Reed writes:

> A little reflection will suggest, however, that our experience of time and its passage is determined by *events,* either external or internal. What the time-gapper is reporting is not that a slice of time has vanished, but that he has failed to register a series of events which would normally have functioned as his time-markers. If he is questioned closely he will admit that his "time-gap" experience did not involve his realization at, say, noon that he had somehow "lost" half an hour. Rather, the experience consists of "waking up" at, say, Florence and realizing that he remembers nothing since Bologna. . . . To understand the experience, however, it is best considered in terms of the absence of *events.* If the time-gapper had taken that particular day off, and spent the morning sitting in his garden undisturbed, he might have remembered just as little of the half-hour in question. He might still describe it in terms of lost time, but he would not find the experience unusual or disturbing. For he would point out that he could not remember what took place between eleven-thirty and twelve simply because nothing of note occurred.

In fact, there is nothing recounted in any of the three works under discussion that cannot be easily explained in terms of normal, though somewhat unusual, psychological behavior we now term *anomalous.* Different and unusual? Yes. Paranormal or otherworldly, requiring the presence of extraterrestrials? No. Diehard proponents may find these explanations unsatisfying, but the open-minded reader will find elaboration and illumination in the textbooks and other works in anomalistic psychology.

Strongly recommended are Reed (1972), Marks and Kammann (1980), Corliss (1982), Zusne and Jones (1982), Radner and Radner (1982), Randi (1982), Gardner (1981), Alcock, (1981), Taylor (1980), and Frazier (1981).

If one looks at the psychodynamics underlying the confabulation of Hopkins's contactees and abductees it is easy to see how even an ordinary, non-FP individual can become one of his case histories. How does Hopkins, for instance, locate such individuals in the first place? Typically, it is done through a selection process; i.e., those individuals who are willing to talk about UFOs—the believers—are selected for further questioning. Those who scoff are summarily dismissed. Once selected for study and permission to volunteer for hypnosis is obtained, a response-anticipation process sets in (Kirsch 1985), and the volunteer is now set up to supply answers to anything that might be asked. Then, during the hypnosis sessions, something similar to the Hawthorne Effect occurs: The volunteer says to himself, "This kindly and famous writer and this important and prestigious doctor are interested in poor little old unimportant me!" And the more the volunteer is observed and interrogated, the greater is the volunteer's motivation to come up with a cracking "good story" that is important and significant and pleasing to these important people. Moreover, as we have long known, it is the perception of reality not the reality itself that is truly significant in determining behavior. If the writer and the doctor-hypnotist are on hand to encourage the volunteer and to suggest to him that his fantasy really happened, who is he to question their interpretation of his experience? Once they tell the contactee how important his fantasy is, he now—if he ever doubted before—begins to believe it himself and to elaborate and embellish it every time it is repeated.

CONSEQUENCES AND SUMMARY

Many readers might feel compelled to ask: "Well, what is so bad about people having fantasies anyway? What harm do they do? You certainly cannot deny they are entertaining. And, as far as the psychiatrists' clients are concerned, whether the fantasies are true or false is of little matter—it's the clients' perceptions of reality that matter and it is this that you have to treat." True, if the client believes it is so, then you have to deal with that belief. The only problem with this lies in its potential for harm. On the national scene today too many lives have been negatively affected and even ruined by well-meaning but tragically misdirected reformers who believe the fantasies of children, the alienated, and the fantasy-prone personality types and have charged innocent people with rape, child molestation, assault, and other sorts of abusive crimes. Nearly every experienced clinician has encountered such claims and then much later has discovered to his chagrin that none of these fantasized events ever happened. Law-enforcement officials are also quite familiar with the products of response expectancies and overactive imaginations in the form of FPs who confess to murders that never happened or to murders that did happen but with which they have no connection. Another problem with the UFO abductee literature is that it is false, misleading, rabble-rousing, sensationalistic, and opportunistically money-grubbing. It takes advantage of people's hopes and fears and diverts them from the literature of science. Our journeys to the stars will be made on spaceships created by determined, hardworking scientists and engineers applying the principles of science, not aboard flying saucers piloted by little gray aliens from some other dimension.

Need we be concerned about an invasion of little gray kidnappers? Amused, yes. Concerned, no.

Should we take Strieber, Hopkins, Kinder, et al. seriously? Not really. They are a long, long way from furnishing reliable and replicable data and their rather shaky hypotheses are miles from anything resembling proof.

Should we insist that such semihysterical and poorly informed journalistic efforts not be published? Only if we all are a bunch of wet blankets and party-poopers. After all, it has been dull lately and these pseudoscientific thrillers have added a welcome note of excitement. And without these works there would be no puzzles to solve. As the old disclaimer says, "It's fun to be fooled, but it's more fun to know!"

Is the human mind a weird and wonderful place and human behavior a billion-ring circus of astounding events? Unquestionably, yes!

One cannot help but be struck by the thought that, in their way, the UFOnaut creations are of some redeeming value. They, besides their value as entertainment, do provide the useful—albeit unintended—service of directing our attention to the extremities of human belief and the perplexing and perennial problem we have in detecting deception. In spite of all our vaunted scientific accomplishments, we have today no absolutely certain, accurate, or reliable means for getting at the truth—for simply determining whether or not someone is lying. Not only are the polygraph and the voice-stress analyzer notoriously unreliable and inaccurate; but the professional interrogators, body-language experts, and psychological testers are also the first to admit their lack of predictive skill. If these abductee claims do no more than stimulate greater efforts toward the development of better "truth detectors," then they will have made an important contribution.

When one man has a private conversation with an angel in the corner, we consider it hallucinatory; when twenty people simultaneously see and talk with this angel, we then have good reason to suspect it may not be hallucinatory. When one man *never* sees an angel in the corner until and unless he is hypnotized, and regressed, even then such reports are not considered hallucinatory. They are merely confabulations. Nor do we classify him as psychologically disturbed or even as lying. He most likely is as normal and mentally healthy as any one of us. If he has been properly primed with powerful suggestions, he may sincerely believe in the truth of his confabulations.

When all things are considered, we shouldn't be too upset with the creators of and believers in what Martin Gardner (1987) calls "the new science-fiction religion." Tolerance is the mark of a civilized mind. We can nevertheless demand that the bookstores and supermarkets classify all such material properly. All UFO, UFO-abductee, past-life, and hypnotic-regression accounts should be taken from the nonfiction counters and moved to the science-fiction shelves.

NOTES

1. People familiar with the unreliability of the polygraph will not be impressed. As for Strieber's sanity, there can be no doubt of this. As *Omni* magazine reported, he received a million-dollar advance from his publisher.

2. The dictionary defines an alien as "one who is strange, wholly different in nature, incongruous. . . ."

REFERENCES

Alcock, James E. 1981. *Parapsychology: Science or Magic?* New York: Pergamon.

AMA Council on Scientific Affairs. 1985. "Scientific Status of Refreshing Recollection by Use of Hypnosis." *Journal AMA* 253, no. 13 (April 5).

Baker, Robert A. 1982. "The Effect of Suggestion on Past-lives Regression." *American Journal of Clinical Hypnosis* 25, no. l: 71–76.

Baker, Robert A., B. Haynes, and B. Patrick. 1983. "Hypnosis, Memory, and Incidental Memory." *American Journal of Clinical Hypnosis* 25, no. 4: 253–562.

Barber, Theodore X. 1969. *Hypnosis: A Scientific Approach.* New York: D. Van Nostrand.

Corliss, William R. 1982. *The Unfathomed Mind: A Handbook of Unusual Mental Phenomena.* New York: Sourcebook.

———. 1983. *Handbook of Unusual Natural Phenomena.* New York: Arlington House.

Curran, Douglas. 1985. *In Advance of the Landing: Folk Concepts of Outer Space.* New York: Abbeville Press.

Frazier, Kendrick, ed. 1981. *Paranormal Borderlands of Science.* Amherst, N.Y.: Prometheus Books.

Gardner, Martin. 1981. *Science: Good, Bad and Bogus.* Amherst, N.Y.: Prometheus Books.

———. 1987. "Science-fantasy Religious Cults." *Free Inquiry* 7, no. 3 (Summer): 31–35.

Gill, M. M., and M. Brenman. 1959. *Hypnosis and Related Slates.* New York: International Universities Press.

Hilgard, Ernest R. 1977. *Divided Consciousness: Multiple Controls in Human Thought and Action.* New York: Wiley.

———. 1981. "Hypnosis Gives Rise to Fantasy and Is Not a Truth Serum." *Skeptical Inquirer* 5, no. 3 (Spring).

Hilgard, Josephine R. 1979. *Personality and Hypnosis: A Study of Imaginative Involvement,* 2d ed. Chicago: University of Chicago Press.

Jung, Carl. 1969. *Flying Saucers: A Modern Myth of Things Seen in the Sky.* Signet Books.

Kirsch, Irving. 1985. "Response Expectancy as a Determinant of Experience and Behavior." *American Psychologist* 40, no. 11: 1189–1202.

Klass, Philip J. 1981. "Hypnosis and UFO Abductions." *Skeptical Inquirer* 5, no. 3 (Spring).

Laurence, Jean-Roch, Robert Nadon, Heather Nogrady, and Campbell Perry. 1986. "Duality, Dissociation, and Memory Creation in Highly Hypnotizable Subjects." *International Journal of Clinical and Experimental Hypnosis* 34, no. 4: 295–310.

Lotus, Elizabeth. 1979. *Eyewitness Testimony.* Cambridge, Mass.: Harvard University Press.

Marks, David, and Richard Kammann. 1980. *The Psychology of the Psychic.* Amherst, N.Y.: Prometheus Books.

McKellar, Peter. 1957. *Imagination and Thinking.* London: Cohen and West.

O'Connell, D. N., R. E. Shor, and M. T. Orne. 1970. "Hypnotic Age Regression: An Empirical and Methodological Analysis." *Journal of Abnormal Psychology Monograph* 76, no. 3: 1–32.

Perry, Campbell, Jean-Roch Laurence, Robert Nadon, and Louise Labelle. 1986. "Past-Lives Regression." In *Hypnosis: Questions and Answers,* edited by Bernie Zilbergeld, M. G. Edelstein, and D. L. Araoz. New York: Norton.

Radner, Daisie, and Michael Radner. 1982. *Science and Unreason.* Belmont, Calif.: Wadsworth.

Randi, James. 1982. *Flim-Flam!* Amherst, N.Y.: Prometheus Books.

Reed, Graham. 1972. *The Psychology of Anomalous Experience.* Boston: Houghton Mifflin.

Sarbin, T. R., and W. C. Coe. 1972. *Hypnosis: A Social Psychological Analysis of Influence Communication.* New York: Holt, Rinehart and Winston.

Sarbin, T. R., and M. L. Andersen. 1967. "Role-theoretical Analysis of Hypnotic Behavior." In *Handbook of Clinical and Experimental Hypnosis,* edited by Jesse E. Gordon. New York: Macmillan.

Sheaffer, Robert. 1981. *The UFO Verdict.* Amherst, N.Y.: Prometheus Books.

Singer, Jerome. 1975. *The Inner World of Daydreaming.* New York: Harper and Row.

Spanos, N. P., and T. X. Barber. 1974. "Toward a Convergence in Hypnotic Research." *American Psychologist* 29, no. 3: 500–11.

Sutcliffe, J. P. 1961. " 'Credulous' and 'Skeptical' Views of Hypnotic Phenomena: Experiments on Esthesia, Hallucinations, and Delusion." *Journal of Abnormal and Social Psychology* 62, no. 2: 189–200.

Tart, Charles, ed. 1969. *Altered States of Consciousness: A Book of Readings.* New York: Wiley.

Taylor, John. 1980. *Science and the Supernatural: An Investigation of Paranormal Phenomena.* New York: Dutton.

True, R. M. 1949. "Experimental Control in Hypnotic Age Regression States." *Science* 110: 583–84.

Wilson, Ian. 1981. *Mind Out of Time.* London: Gollancz.

Wilson, Sheryl C., and T. X. Barber. 1983. "The Fantasy-prone Personality: Implications for Understanding Imagery, Hypnosis, and Parapsychological Phenomena." In *Imagery: Current Theory, Research and Application,* edited by A. A. Sheikh. New York: Wiley.

Zusne, Leonard, and Warren H. Jones. 1982. *Anomalistic Psychology: A Study of Extraordinary Phenomena of Behavior and Experience.* Hillsdale, N.Y.: Erlbaum.

30. DIAGNOSES OF ALIEN KIDNAPPINGS THAT RESULT FROM CONJUNCTION EFFECTS IN MEMORY

Robyn M. Dawes and Matthew Mulford

Events and feelings may be better recalled when they occur in combination than singly, to the point that a conjunction of two alleged events or feelings may be judged to have occurred with greater frequency in one's life than one of them alone. One part of a conjunction can facilitate recall of the conjunction, and hence of another part of the experience—and combinations of events can be judged to be more probable than their components (Tversky and Kahneman 1983). The observer to whom it is reported, however, knows that such a conjunction is necessarily less probable than any one of its components. Thus, the observer may attach special significance to such a conjunction.

For example, in supporting a conclusion that post-traumatic stress from kidnapping by aliens is a major mental-health problem in this country (allegedly affecting at least 2 percent of the population), Hopkins and Jacobs (1992) cite the rate of affirmative responses to a recent Roper Poll question: "How often has this happened to you: Waking up paralyzed with a sense of a strange person or presence or something else in the room?" Their rationale for considering affirmative responses particularly diagnostic of alien kidnapping involves the conjunction of the two components in the question: "A fleeting sensation of paralysis is not unusual in either hypnogogic or hypnopompic states, but adding the phrase 'with a sense of a strange person or presence in the room' forcefully narrows the scope of the question" (p. 56).

As part of a (much) larger study, we asked that same Roper Poll ques-

This article originally appeared in the *Skeptical Inquirer* 18 (Fall 1993). Reprinted with permission.

TABLE 1: RESPONSE FREQUENCIES

How often has this happened to you: Waking up paralyzed with a sense of a strange person or presence or something else in the room?[1]

Has not happened	Has happened once or twice	Has happened more than twice
(*N*)	(*N*)	(*N*)
87	36	21

How often has this happened to you: Waking up paralyzed?

Has not happened	Has happened once or twice	Has happened more than twice
(*N*)	(*N*)	(*N*)
124	12	8

[1]Exact wording from the Roper Poll.

tion of 144 subjects (mainly University of Oregon students and some townspeople interested in the $20 pay for two hours). Forty percent answered that this had happened to them at least once. A randomly selected control group of 144 subjects in the same study were asked simply how often they remembered waking up paralyzed. Only 14 percent answered that this had happened to them at least once, (chi-square = 24.26; $p < .001$, phi = .29). (See table 1.) The contingency was stronger for women (phi = .44) than for men (phi = .17), significantly so according to a Goodman-Plackett chi-square value of 4.74. Nevertheless it was significant for both sexes (with chi-squares of 25.38 and 4.43, respectively).

Thus, due to a conjunction effect in memory, the added phrase "with a sense of a strange person or presence . . . in the room" actually "*broadens* the scope" of the question, rather than narrowing it. Hopkins and Jacobs are, of course, correct in maintaining that the additional phrase *should* "narrow the scope." It's just that the phrase doesn't. What they have discovered, therefore, is evidence not of alien kidnappings, but of a common irrationality in the way we recall our lives.

REFERENCES

Hopkins, B., and D. M. Jacobs. 1992. "How This Survey Was Designed." In Bigelow Holding Company, *An Analysis of the Data from Three Major Surveys Conducted by the Roper Organization,* 55–58.

Tversky, A., and D. Kahneman. 1983. "Extensional versus Intuitive Reasoning: The Conjunction Fallacy in Probability Judgments." *Psychological Review* 90: 293–315.

31. STUDYING THE PSYCHOLOGY OF THE UFO EXPERIENCE

Robert A. Baker

Empirical studies of people alleged to have had UFO experiences are hard to come by because of the relative rarity of such claimants, the shortage of interested and scientifically trained investigators, and the esoteric nature of the subject matter. A new study by Nicholas P. Spanos and colleagues in the Carleton University Department of Psychology in Ottawa, Canada, is noteworthy in that it has managed to overcome these obstacles ("Close Encounters: An Examination of UFO Experiences," by Nicholas P. Spanos, Patricia A. Cross, Kirby Dickson, and Susan C. DuBreuil, *Journal of Abnormal Psychology* 102, no. 4: 624–32, 1993). This study is significant, in my opinion, because it shows that the earlier explanatory accounts of this phenomenon are somewhat flawed and it also points to the most probable or "true" explanation for these curious reports.

Recognizing at the outset that claims of alien encounters and alien abductions are a social-psychological rather than a material or physical problem, Spanos et al. sought to test two general hypotheses. The first is that people making such claims are psychologically or psychosocially disturbed, i.e., pathological. The second argues that these people have strong imaginative powers and are "fantasy prone," i.e., people who are easily hypnotized and who weave elaborate fantasies around internal sensations and external suggestions and then are convinced that the fantasies are "real."

Using 176 adults, one group of 49 UFO reporters recruited through newspaper ads and two comparison groups—one of 53 community vol-

This article originally appeared in the *Skeptical Inquirer* 18 (Spring 1994). Reprinted with permission.

unteers and the other of 74 university students—the psychologists gave the three groups a series of psychological tests and questionnaires. Subjects in the UFO group also received an hour-long, semistructured interview and were asked about their UFO experiences and their beliefs. The interviews were scored by independent judges and were rated for the emotional intensity and subjective reality of the UFO encounter.

The psychological measures used are all standard, valid, and reliable. Well-accepted measurements were used to determine psychopathology, intelligence, social and emotional adjustment, cognitive ability, and fantasy proneness.

In analyzing the results, the UFO reporters were divided into two groups: those who simply saw lights in the sky and scored very low on the belief intensity dimension and those who scored much higher on the emotional and reality scales. All four groups were then compared: 31 UFO-intense subjects; 18 UFO-nonintense subjects; 53 community adults; and 74 students. Differences between the groups on each of the test battery variables were analyzed with a series of one-way analyses of variance.

Both UFO groups held significantly more exotic beliefs than subjects in the comparison groups, and they also attained higher scores on five of the psychological health variables. In short, the findings provide no support whatsoever for the hypothesis that UFO reporters are psychologically pathological. Regarding the "fantasy" hypothesis, no group differences were found on the temporal lobe lability scale, on the three imaginal propensity measures, and on three indexes of hypnotizability—all recognized measures of fantasy proneness. There were, however, major differences between the intense and nonintense UFO groups with regard to whether or not their experiences were sleep related. UFO-intense subjects reported that their experiences were sleep related significantly more often than did the nonintense subjects. Only one subject in the nonintense group said his experience was related to falling asleep, dreaming, or waking up; 58 percent of the intense subjects described their experience as sleep related.

In their discussion, the authors interpret their findings as failing to confirm either the psychopathology or fantasy-prone hypothesis about UFO close-encounter experiences. Not only were the UFO groups as mentally healthy as the controls but the UFO groups were no different from the controls on any of the imaginal propensity measures. The finding that did most clearly differentiate the UFO and comparison groups was belief in UFOs (as alien spacecraft) and in the reality of alien life-forms. The authors also suggest that many of the UFO subjects believed in alien existence long before having any UFO experiences. In their words (p. 631):

> Our findings suggest that intense UFO experiences are more likely to occur in individuals who are disposed to esoteric beliefs in general and alien beliefs in particular and who interpret unusual sensory and imaginal experiences in terms of the alien hypothesis. Among UFO believers, those with stronger propensities toward fantasy production were particularly likely to generate such experiences. Moreover such experiences were likely to be generated and interpreted as real events rather than imaginings when they were associated with restricted sensory environments that contributed to confusion between internally produced images and sensations and external events (e.g., experiences that occurred at night and in association with sleep).

The authors also caution that to understand the elaborate and often bizarre reports of those in the intense group it is important to note that most of these reports (60 percent) were sleep related. Although some were merely dreams of UFOs and aliens, the sleep-related UFO experiences involving paralysis were also usually accompanied by visual or auditory hallucinations, or both, and sometimes by the sense of a presence that was somehow felt but not seen.

Baker (1992), in an unpublished but widely distributed paper, "Alien Abductions or Human Productions: Some Not So Unusual Personal Experiences," has also called attention to the striking similarity between reports of alien abductions (as chronicled by clients of Hopkins [1987], Jacobs [1992], and Mack [1992], as well as Whitley Strieber, regarding his own abduction claim in *Communion*) and the medical reports from people having sleep paralysis with hypnagogic and hypnopompic hallucinations (Rehm 1991; Siegel 1992). For all practical purposes, the two sets of reports are identical. A comparison of UFO-abduction reports and the medical reports of sleep paralysis with accompanying hypnagogic and hypnopompic hallucinations show the two to differ only in the hallucinatory content. Ghosts and demons dominate the medical, historical, and folklore narrations; UFOs and aliens appear in modern reports. Explanations for other psychological manifestations alleged to accompany the abduction scenario are also provided in the Baker report.

While the Spanos et al. study adds significantly to our understanding of the personality characteristics of those individuals reporting UFO encounters and who believe in the extraterrestrial hypothesis (which is, after all, the crux of the matter), it should not be forgotten that people claiming abduction by aliens are relatively few in number. This is certainly the case in the Spanos study. Most important, the Spanos study outcome must not under any circumstances be construed as providing aid and comfort to those believers in either the extraterrestrial hypothesis or the credibility of alien abductions. This has, unfortunately, already happened in several instances where the media have left the impression that if those who believe in UFOs and alien abductions are neither "crazy" nor "fantasy prone" then their beliefs in extraterrestrial visitations have been validated.

The major importance of this study is that it shifts attention from the false leads of pathology and neurological aberrations to the true path of false beliefs that are engendered by a common but little recognized type of sleep disorder.

The fact that UFO contactees and abductees are not pathological is, however, not surprising to experienced anomalistic psychologists, who have long been aware that delusions, i.e., false beliefs, are common coin and in no way are indicative of pathology.

Unless our delusions interfere with our social and personal adjustment or are so pervasive as to endanger our lives and the lives of others, they are usually ignored. Often these false beliefs may serve to help us keep our mental balance and even cope with life's stresses and strains, as in the case of our religions. Many noted and distinguished psychiatrists also hold to what most scientifically trained people would consider to be false beliefs. Raymond Moody, in his latest book, *Reunion* (1993), says he converses frequently with visions of his dead relatives. M. Scott Peck believes in demon possession as the true explanation for many multiple personality disorders (1983, 1984) and Colin Ross believes that most skilled athletes have and use psychokinetic powers (1989). According to the December 27, 1993, *Time* magazine, 69 percent of the American public believe in the existence of angels and 46 percent believe they have their own guardian angel. One person's conviction and reality are another person's fantasy and delusion. Some people believe they have been abducted by aliens. Most of us do not.

It is also important for the public to realize that people who *are* pathological or "crazy" are not "crazy" all over or crazy all the time. We are only "crazy" in spots and only on occasion, not continuously. Psychiatric diagnosis is not an easy task, as the *Diagnostic and Statistical Manual of Mental Disorders,* the famous Rosenhan study (1973), and the difficulties of dealing with factitious disorders (Feldman and Ford 1994) attest. In this regard the Spanos et al. conclusion, that "among UFO believers, those with strong propensities toward fantasy production were particularly likely to generate such experiences" (p. 631), again calls attention to the important role that imagination and fantasy play in the production of alien-contact and alien-abduction reports. A similar conclusion was also reached by Ring and Rosing (1990) in their survey of people reporting UFO encounters and abductions.

Although the two studies are somewhat similar in intent and design, the Spanos et al. study is by far the superior one and sticks to the mundane three-dimensional world of cold, hard reality rather than soaring off into alternate realities and universes to account for a very worldly human experience based upon beliefs in alien visitation. These beliefs, say the authors, "serve as templates against which people shape ambiguous external information, diffuse physical sensations, *and vivid imaginings* [emphasis added] into alien encounters that are experienced as real events" (p. 631). Of the experimental and empirical studies we have of this particular problem, the Spanos et al. study is the best thus far.

REFERENCES

Baker, Robert A. 1992. "Alien Abductions or Human Productions: Some Not So Unusual Personal Experiences." Unpublished manuscript. Available upon request from the author.

Feldman, Marc D., and C. V. Ford. 1994. *Patient or Pretender.* New York: John Wiley.

Gibbs, Nancy. 1993. "Angels among Us." *Time,* December 27, 1993, 56–65.

Hopkins, Budd. 1987. *Intruders: The Incredible Visitations At Copley Woods.* New York: Random House.

Jacobs, David. 1992. *Secret Life: Firsthand Accounts of UFO Abductions.* New York: Simon and Schuster.

Mack, John E. 1992. "Helping Abductees." *International UFO Reporter,* July/August, 10–20.

Moody, Raymond. 1993. *Reunions: Visionary Encounters with Departed Loved Ones.* New York: Villard Books, Random House.

Peck, M. Scott. 1983. *People of the Lie.* New York: Simon and Schuster.

———. 1984. "A Psychiatrist's View of Exorcism." *Fate* 37, no. 9: 87–96.

Rehm, Stanley R. 1991. "Sleep Paralysis and Nocturnal Dyspnea." Paper presented at Advances in Pulmonary and Critical Care Medicine International Symposium, Vienna, Austria, August 29–30, 1991, University of Kentucky Medical Center, Lexington, Ky.

Ring, Kenneth, and C. Rosing. 1990. "The Omega Project: A Psychological Survey of Persons Reporting Abductions and Other UFO Encounters." *Journal of UFO Studies* 2: 59–98.

Rosenhan, D. L. 1973. "Being Sane in Insane Places." *Science* 179: 250–58.

Ross, Colin A. 1989. *Multiple Personality Disorder: Diagnosis, Clinical Features, and Treatment* (specifically pp. 185–88). New York: John Wiley.

Siegel, Ronald K. 1992. *Fire in the Brain: Clinical Tales of Hallucination.* New York: Dutton.

32. *TIME* CHALLENGES JOHN MACK'S UFO ABDUCTION EFFORTS

Philip J. Klass

The widely read "Outlook" section in the Sunday, April 17, *Washington Post* carried a front-page feature article by Harvard psychiatrist John E. Mack with the headline: "ALIEN RECKONING: Many Americans Claim They've Been Abducted by Extraterrestrials. A Once-Skeptical Harvard Psychiatrist Believes Them." If any of the *Post*'s top editorial officials believe there is even one chance in a million that UFO abductions are really occurring, one should expect that they would have assigned a team of investigative reporters to look into Mack's claims—*before publishing his article.* The *Post* could have assigned the story to its Robert Woodward, who played a key role in breaking the Watergate coverup.

Time magazine *did* assign an investigative reporter—James Willwerth—to look into Mack's claims. Less than a week after publication of the *Post* article, *Time*'s April 25 issue provided its readers with a two-page critical analysis of Mack's modus operandi, titled: "The Man from Outer Space."

The *Time* article revealed that one of the "abductees" Mack had worked with—Donna Bassett—was really a researcher who had infiltrated his "support group." According to *Time,* after reading books and articles on UFO abductions Bassett was able to "recall under hypnosis" for Mack her childhood experiences with an ET friend "named Jane, who healed her hands after a neighbor stuck them in boiling fudge to punish her for snooping."

This article originally appeared in the *Skeptical Inquirer* 18 (Summer 1994). Reprinted with permission.

At one of the sessions, which took place in a bedroom of Mack's house, Bassett told Mack she "recalled" being taken aboard a flying saucer during the Cuban missile crisis, where she saw President Kennedy and Premier Khrushchev, who was crying. She told Mack that, to comfort Khrushchev, she sat on his lap, put her arms around his neck, and reassured him that the crisis would be resolved. "Hearing her tale, Mack became so excited that he leaned on the bed too heavily, and it collapsed," according to *Time.*

When *Time*'s Willwerth revealed to Mack that Bassett had concocted her abduction tales, "Mack declined to discuss her case, though he hinted that he had doubts about her reliability."

The *Time* article concludes: "Mack's view of the UFO phenomenon reflects a larger philosophical stance that rejects 'rational' scientific explanations and embraces a hazier New Age reality. 'I don't know why there's such a zeal to find a conventional physical explanation,' he says. 'I don't know why people have such trouble simply accepting the fact that something unusual is going on here. . . . We've lost all that ability to know a world beyond the physical. . . . I am a bridge between those two worlds.' "

With Mack's widespread appearances on major TV and radio talk-shows to promote his new book, *Abduction: Human Encounters with Aliens,* Mack is displacing his former mentor, Budd Hopkins, as the chief guru of the UFO-abduction cult. Because Mack is a psychiatrist who formerly headed Harvard's department of psychiatry, his claims carry more weight than those of Hopkins, who is a professional artist, or Hopkins's principal deputy, David Jacobs, a history professor at Temple University.

Although Mack credits Hopkins with having introduced him to the UFO-abduction field, and each publicly refers to the other as "my good friend," their relations have deteriorated recently because of their basically different philosophical views on UFO abductions. Hopkins and Jacobs, understandably, view UFO abductions as evil, traumatic events—comparable to rape.

Mack concedes that abduction victims suffer what he calls "ontological shock" as "the reality of their encounters sinks in." However, as Mack explained in his *Post* article, the net result "is the discovery of a new and altered sense of their place in the cosmic design, one that is more modest, respectful and harmonious in relation to Earth and its living systems. A heightened sense of the sacredness of the natural world is experienced along with deep sadness about the apparent hopelessness of Earth's environmental crisis." In other words, in the long run Mack seemingly believes that UFO abductions are beneficial to the "experiencers," as he calls them.

33. A STUDY OF FANTASY PRONENESS IN THE THIRTEEN CASES OF ALLEGED ENCOUNTERS IN JOHN MACK'S ABDUCTION

Joe Nickell

INTRODUCTION

Since Robert A. Baker's pioneering article [see chapter 29 in this volume] appeared in the *Skeptical Inquirer* (Baker 1987–1988), a controversy has raged over his suggestion that self-proclaimed "alien abductees" exhibited an array of unusual traits that indicated they had fantasy-prone personalities. Baker cited the "important but much neglected" work of Wilson and Barber (1983), who listed certain identifying characteristics of people who fantasize profoundly. Baker applied Wilson and Barber's findings to the alien-abduction phenomenon and found a strong correlation. Baker explained how a cursory examination by a psychologist or psychiatrist might find an "abductee" to be perfectly normal, while more detailed knowledge about the person's background and habits would reveal to such a trained observer a pattern of fantasy proneness.

For example, Baker found Whitley Strieber—author of *Communion,* which tells the "true story' of Strieber's own alleged abduction—to be "a classic example of the [fantasy-prone personality] genre." Baker noted that Strieber exhibited such symptoms as being easily hypnotized, having vivid memories, and experiencing hypnopompic hallucinations (i.e. "waking dreams"), as well as being "a writer of occult and highly imaginative novels" and exhibiting other characteristics of fantasy proneness. A subsequent, but apparently independent, study by Bartholomew and Basterfield (1988) drew similar conclusions.

This article originally appeared in the *Skeptical Inquirer* 20 (May/June 1996). Reprinted with permission.

Wilson and Barber's study did not deal with the abduction phenomenon (which at the time consisted of only a handful of reported cases), and some of their criteria seem less applicable to abduction cases than to other types of reported phenomena, such as psychic experiences. Nevertheless, although the criteria for fantasy proneness have not been exactly codified, they generally include such features as having a rich fantasy life, showing high hypnotic susceptibility, claiming psychic abilities and healing powers, reporting out-of-body experiences and vivid or "waking" dreams, having apparitional experiences and religious visions, and exhibiting automatic writing. In one study, Bartholomew, Basterfield, and Howard (1991) found that, of 152 otherwise normal, functional individuals who reported they had been abducted or had persistent contacts 'with extraterrestrials, 132 had one or more major characteristics of fantasy-prone personality.

Somewhat equivocal results were obtained by Spanos et al. (1993), although their "findings suggest that intense UFO experiences are more likely to occur in individuals who are predisposed toward esoteric beliefs in general and alien beliefs in particular and who interpret unusual sensory and imagined experiences in terms of the alien hypothesis. Among UFO believers, those with stronger propensities toward fantasy production were particularly likely to generate such experiences" (Spanos et al. 1993, 631).

A totally dismissive view of these attempts to find conventional psychological explanations for the abduction experience is found in the introduction to psychiatrist John Mack's *Abduction: Human Encounters with Aliens* (1994). Mack states unequivocally: "The effort to discover a personality type associated with abductions has also not been successful." According to Mack, since some alleged abductions have reportedly taken place in infancy or early childhood, "Cause and effect in the relationship of abduction experiences to building of personality are thus virtually impossible to sort out" (Mack 1994, 5). But surely it is Mack's burden to prove his own thesis that the alien hypothesis does have a basis in fact beyond mere allegation. Otherwise the evidence may well be explained by a simpler hypothesis, such as the possibility that most "abductees" are fantasy-prone personality types. (Such people have traits that cut across many different personality dimensions; thus conventional personality tests are useless for identifying easily hypnotizable people. Some "abductees" who are not fantasy prone may be hoaxers, for example, or exhibit other distinctive personality traits or psychological problems.) Mack's approach to the diagnosis and treatment of his "abductee" patients has been criticized by many of his colleagues (e.g., Cone 1994).

METHODOLOGY

To test the fantasy-proneness hypothesis, I carefully reviewed the thirteen chapter-length cases in Mack's *Abduction* (chapters 3–15), selected from the forty-nine patients he most carefully studied out of seventy-six "abductees." Since his presentation was not intended to include fantasy proneness, certain potential indicators of that personality type—like a subject's having an imaginary playmate—would not be expected to be present. Nevertheless, Mack's rendering of each personality in light of the person's alleged abduction experiences was sufficiently detailed to allow the extraction of data pertaining to several indicators of fantasy proneness. They are the following:

1. *Susceptibility to hypnosis.* Wilson and Barber rated "hypnotizability" as one of the main indicators of fantasy proneness. In all cases, Mack repeatedly hypnotized the subjects without reporting the least difficulty in doing so. Also, under hypnosis the subjects did not merely "recall" their alleged abduction experiences but all of them *reexperienced* and *relived* them in a manner typical of fantasy proneness (Wilson and Barber 1983, 373–79). For example, Mack's patient "Scott" (No. 3) was so alarmed at "remembering" his first abduction (in a pre-Mack hypnosis session with another psychiatrist) that, he said, "I jumped clear off the couch" (Mack 1994, 81); "Jerry" (No. 4) "expressed shock over how vividly she had relived the abduction," said Mack (1994, 112); similarly, "Catherine" (No. 5) "began to relive" a feeling of numbness and began "to sob and pant" (Mack 1994, 140).

2. *Paraidentity.* I have used this term to refer to a subject's having had imaginary companions as a child (Wilson and Barber 1983, 346–47) and/or by extension to claiming to have lived past lives or to have a dual identity of some type. Of their fantasy-prone subjects, Wilson and Barber stated: "In fantasy they can do anything—experience a previous lifetime, experience their own birth, go off into the future, go into space, and so on." As well, "While they are pretending, they become totally absorbed in the character and tend to lose awareness of their true identity" (Wilson and Barber 1983, 353, 354).

Thus, as a child, "Ed" (No. 1) stated: "Things talked to me. The animals, the spirits. . . . I can sense the earth" (Mack 1994, 47); "Jerry" (No. 4) said he has had a relationship with a tall extraterrestrial being since age five (Mack 1994, 113). At least four of Mack's subjects (Nos. 5, 7, 9, and 10) said they have had past-life experiences (160–62, 200, 248, 259), and seven (Nos. 3, 6, 7, 8, 9, 11, and 12) said they have some sort of dual identity (92–93, 173, 200, 209, 243, 297, and 355–56). For example "Dave"

(No. 10) said he considers himself "a modern-day Indian"; while "Peter" (No. 11) under hypnosis said he becomes an alien and speaks in robotic tones (Mack 1994, 275, 277, 297). In all, eleven of Mack's thirteen featured subjects exhibited paraidentity.

3. *Psychic experiences.* Another strong characteristic of fantasy proneness according to Wilson and Barber (1983, 359–60) is that of having telepathic, precognitive, or other types of psychic experience.

One hundred percent of Mack's thirteen subjects claimed to have experienced one or more types of alleged psychical phenomena, most reporting telepathic contact with extraterrestrials.

"Catherine" (No. 5) also claimed she can "feel people's auras"; "Eva" (No. 9) said she is able to perceive beyond the range of the five senses; and "Carlos" (No. 12) said he has had "a history of what he calls 'visionary' experiences" (Mack 1994, 157, 245, 332).

4. *"Floating" or out-of-body experiences.* Wilson and Barber (1983, 360) stated: "The overwhelming majority of subjects (88 percent) in the fantasy-prone group, as contrasted to few (8 percent) in the comparison group, report realistic out-of-the-body experiences" (which one subject described as "a weightless, floating sensation" and another called "astral travel"). Only one of Mack's thirteen subjects (No. 2) failed to report this; of the other twelve, most described, under hypnosis, being "floated" from their beds to an awaiting spaceship. Some said they were even able to drift through a solid door or wall, that being a further indication of the fantasy nature of the experience (more on this later). Also, "Eva" (No. 9) stated that she had once put her head down to nap at her desk and then "saw myself floating from the ceiling. . . . My consciousness was up there. My physical body was down there" (Mack 1994, 237). Also, in the case of "Carlos" (No. 12), "Flying is a recurring motif in some of his more vivid dreams" (Mack 1994, 338).

5. *Vivid or "waking" dreams, visions, or hallucinations.* A majority of Wilson and Barber's subjects (64 percent) reported they frequently experienced a type of dream that is particularly vivid and realistic (Wilson and Barber 1983, 364). Technically termed *hypnogogic or hypnopompic hallucinations* (depending on whether they occur, respectively, while the person is going to sleep or waking), they are more popularly known as "waking dreams" or, in earlier times, as "night terrors" (Nickell 1995, 41). Wilson and Barber (1983, 364) reported that several of their subjects "were especially grateful to learn that the 'monsters' they saw nightly when they were children could be discussed in terms of 'what the mind does when it is nearly, but not quite, asleep.' " Some of Wilson and Barber's subjects (six in the fantasy-prone group of twenty-seven, contrasted with none in the comparison group of twenty-five) also had religious

visions, and some had outright hallucinations (Wilson and Barber 1983, 362–63, 364–65, 367–71).

Of Mack's thirteen selected cases, all but one (No. 13) reported either some type of especially vivid dream, or vision, or hallucination. For example, "Scott" (No. 3) said he had "visual hallucinations" from age twelve; "Jerry" (No. 4) recorded in her journal "vivid dreams of UFOs" as well as "visions"; and "Carlos" (No. 12) had the previously mentioned "visionary" experiences and dreams of flying (Mack 1994, 82, 112). Almost all of Mack's subjects (Nos. 1–11), like "Sheila" (No. 2), had vivid dreams with strong indications of hypnogogic/hypnopompic hallucination (Mack 1994, 38, 56, 80, 106, 132, 168–69, 196, 213, 235, 265–67, 289).

6. *Hypnotically generated apparitions.* Encountering apparitions (which Wilson and Barber define rather narrowly as "ghosts" or "spirits") is another Wilson-Barber characteristic (contrasted with only sixteen percent of their comparison group). A large number of the fantasizers also reported seeing classic hypnogogic imagery, which included such apparitionlike entities as "demon-type beings, goblins, gargoyles, monsters that seemed to be from outer space" (Wilson and Barber 1983, 364).

Mack's subjects had a variety of such encounters, both in their apparent "waking dreams" and under hypnosis. Only the latter were considered here; all thirteen subjects reported seeing one or more types of outer-space creatures during hypnosis.

7. *Receipt of special messages.* Fifty percent of Wilson and Barber's fantasizers (contrasted with only 8 percent of their comparison subjects) reported having felt that some spirit or higher intelligence was using them "to write a poem, song, or message" (Wilson and Barber 1983, 361).

Of Mack's thirteen abductees, all but one clearly exhibited this characteristic, usually in the form of receiving telepathic messages from the extraterrestrials and usually with a message similar to the one given "Arthur" (No. 13) "about the danger facing the earth's ecology" (Mack 1994, 381). Interestingly, many of these messages just happen to echo Mack's own apocalyptic notions (e.g., 3, 412), indicating Mack may be leading his witnesses.

In the case of "Eva" (No. 9), the aliens, who represented a "higher communication" (Mack 1994, 243, 247), purportedly spoke through her and described her "global mission." "Jerry" (No. 4) produced a "flood of poetry," yet stated, "I don't know where it's coming from" (99); "Sara" (No. 7) has been "spontaneously making drawings with a pen in each hand [of aliens]" although she had never used her left hand before; and "Peter" (No. 11) stated he has "always known that I could commune with God" and that the aliens "want to see if I'm a worthy leader" (Mack 1994, 99, 192, 288, 297).

FIGURE 1. ALIEN ENCOUNTER CASES FROM JOHN MACK'S *ABDUCTION* STUDIED FOR FANTASY PRONENESS

	Case Number from Mack's *Abduction*												
Fantasy Proneness Markers	1	2	3	4	5	6	7	8	9	10	11	12	13
1. Susceptibility to Hypnosis	•	•	•	•	•	•	•	•	•	•	•	•	•
2. Paraidentity (Imaginary Companions, Past Lives, Dual Identities)	•		•	•	•	•	•	•	•	•	•	•	
3. Psychic Experiences	•	•	•	•	•	•	•	•	•	•	•	•	•
4. 'Floating' or Out-of-Body Experiences	•		•	•	•	•	•	•	•	•	•	•	•
5. Vivid or 'Waking' Dreams/ Visions/Hallucinations	•	•	•	•	•	•	•	•	•	•	•	•	
6. Hypnotically Generated Apparitions	•	•	•	•	•	•	•	•	•	•	•	•	•
7. Receipt of Special Messages	•		•	•	•	•	•	•	•	•	•	•	•

RESULTS

One of Mack's subjects ("Sheila," No. 2) exhibited four of the seven fantasy-prone indicators, and another (Arthur, No. 13) exhibited five; the rest showed all seven characteristics. These results are displayed in figure 1.

Although not included here, *healing*—that is, the subjects' feeling that they have the ability to heal—is another characteristic of the fantasy-prone personality noted by Wilson and Barber (1983, 363). At least six of Mack's thirteen subjects exhibited this. Other traits, not discussed by Wilson and Barber but nevertheless of possible interest, are the following (together with the number of Mack's thirteen subjects that exhibit it): having seen UFOs (9); New Age or mystical involvement (11); Roman Catholic upbringing (six of nine whose religion was known or could be inferred); previously being in a religio-philosophical limbo/quest for meaning in life (10); and involvement in the arts as a vocation or avocation (5). For example, while apparently neither an artist, healer, or UFO sighter, "Ed" (No. 1) had "a traditional Roman Catholic upbringing" and—as rather a loner who said he felt "lost in the desert"—he not only feels he can "talk to plants" but said he has "practiced meditation and studied Eastern philosophy in his struggle to find his authentic path" (Mack 1994, 39, 41–42). "Carlos" (No. 12) is an artist/writer/"fine arts professor" involved in theatrical production who said he has seen UFOs and has a "capacity as a healer"; raised a Roman Catholic, and interested in numerology and mythology, he calls himself "a shaman/artist teacher" (Mack 1994, 330, 332, 340–41, 357).

Also of interest, I think, is the evidence that many of Mack's subjects fantasized while under hypnosis. For example—in addition to aliens—"Ed" (No. 1) also said he saw earth spirits whom he described as "mirthful and, playful creatures" (48); and "Joe" (No. 6) said he saw "mythic gods, and winged horses." "Joe" also "remembered" being born (Mack 1994, 170,184). "Catherine" (No. 5), "Sari" (No. 7), "Paul" (No. 8), and "Eva," (No. 9) said they had past-life experiences or engaged in time-travel while under hypnosis. Several said they were able to drift through solid doors or walls, including "Ed" (No. 1), "Jerry' (No. 4), "Catherine" (No. 5), "Paul" (No. 8), "Dave" (No. 10), and "Arthur" (No. 13). "Carlos" (No. 12) claimed his body was transmuted into light. I have already mentioned that under hypnosis "Peter" (No. 11) said he becomes an alien and speaks in an imitative, robotic voice. In all, eleven of Mack's thirteen subjects (all but Nos. 2 and 3) appear to fantasize under hypnosis. Of course it may be argued that there really are "earth spirits" and "winged horses," or that the extraterrestrials may truly have the ability to time travel or dematerialize bodies, or that any of the other examples I have given as evidence of fantasizing are really true. However, once again the burden of proof is on the claimant and until that burden is met, the examples can be taken as further evidence of the subjects' ability to fantasize.

CONCLUSIONS

Despite John Mack's denial, the results of my study of his best thirteen cases show high fantasy proneness among his selected subjects. Whether or not the same results would be obtained with his additional subjects remains to be seen. Nevertheless, my study does support the earlier opinions of Baker and of Bartholomew and Basterfield that alleged alien abductees tend to be fantasy-prone personalities. Certainly, that is the evidence for the very best cases selected by a major advocate.

NOTE

I am grateful to psychologists Robert A. Baker and Barry Beyerstein for reading this study and making helpful suggestions.

REFERENCES

Baker, Robert A. 1987–1988. "The Aliens among Us: Hypnotic Regression Revisited." *Skeptical Inquirer* 12, no. 2 (Winter): 147–62.

Bartholomew, Robert E., and Keith Basterfield. 1988. "Abduction States of Consciousness." *International UFO Reporter,* March/April.

Bartholomew, Robert E., Keith Basterfield, and George S. Howard, 1991. "UFO Abductees and Contactees: Psychopathology or Fantasy Proneness?" *Professional Psychology: Research and Practice* 22, no. 3: 215–22.

Cone, William. 1994. "Research Therapy Methods Questioned." *UFO* 9, no. 5: 32–34.

Mack, John. 1994. *Abduction: Human Encounters with Aliens.* New York: Simon and Schuster.

Nickell, Joe. 1995. *Entities: Angels, Spirits, Demons, and Other Alien Beings.* Amherst, N.Y.: Prometheus Books.

Spanos, Nicholas P, Patricia A. Cross, Kirby Dickson, and Susan C. DuBreuil. 1993. "Close Encounters: An Examination of UFO Experiences." *Journal of Abnormal Psychology* 102, no. 4: 624–32.

Wilson, Sheryl C., and Theodore X. Barber. 1983. "The Fantasy-prone Personality: Implications for Understanding Imagery, Hypnosis, and Parapsychological Phenomena." In *Imagery, Current Theory, Research and Application,* edited by Anees A. Sheikh, 340–90. New York: Wiley.

34. NOVA'S ALIEN ABDUCTION PROGRAM SHOWS QUESTIONABLE TECHNIQUES

C. Eugene Emery, Jr.

There's a maxim in journalism that showing is better than telling. Instead of stating that someone is a crook or a saint, showing them doing crooked or saintly things will leave a far more lasting impression.

PBS's February 27, 1996, "Nova" program on alien abductions ("Kidnapped by UFOs?" written, produced, and directed by Denise Dilanni) tried to follow that rule, and the result was one of the best, most authoritative television programs on alien abductions produced to date.

The idea that regular folks have been held captive by space aliens, used as guinea pigs, and served as involuntary donors of eggs and sperm to produce a hybrid of human and extraterrestrial is a compelling piece of folklore.

And because the human brain seems programmed to give more weight to one well-told story than to piles of data suggesting that the story is false, the similar-sounding tales told by UFO abductees have compelled a lot of people to believe that investigators like psychiatrist John Mack or artist Budd Hopkins really are dealing with the victims of alien kidnappings.

Unfortunately for skeptics trying to lend a credible counterpoint to this scenario, the science behind the examination of UFO claims can be subtle and sensitive, as delicate as pointing out that a therapist may be suggesting to hypnotized persons that they might want to interpret their dream as an abduction experience.

But ours is not an age of subtlety. The tabloid talk and news shows,

---

This article originally appeared in the *Skeptical Inquirer* 20 (May/June 1996). Reprinted with permission.

where this phenomenon has largely played itself out, want issues cast in the harsh contrast of black and white, right and wrong. They want tales of legendary journeys, shocking victimization, coverup or ineptitude by authority figures. The UFO abduction tales have all the right ingredients.

On these television programs, if a skeptic gets the opportunity to raise the possibility that the hypnotist was shaping the recollections, the hypnotist and the subject roundly deny such influence. If the skeptic (and there's seldom more than one on these shows) questions whether the experiences are real, the UFO proponents brandish research suggesting that abductees suffer from no mental illness and argue, in effect, that because these people are not crazy, their experiences are real.

"Nova" tried to explore those delicate issues by taking viewers to hypnosis sessions held by Hopkins, showing Hopkins interviewing two children for the first time; and letting the public hear what happens when an abductee claims to have been aboard a UFO with John Kennedy and Nikita Khrushchev at the height of the Cuban missile crisis.

Hopkins's visit with the two children was particularly revealing, showing how the man who popularized current UFO abduction folklore won't take no for an answer.

When he shows four-year-old Ryan a stereotypical drawing of a big-eyed space alien and asks if he recognizes the picture, the boy shakes his head. Nonetheless, Hopkins asks him to make up a story about the creature in the drawing, in which Hopkins finds elements that suggest a kidnapping.

When he turns to Ryan's younger sister, toddler Paula, with the same picture, Hopkins asks, "Is he a nice guy or a bad guy?"

"Bad guy," Paula answers.

"Do you like him?" Hopkins asks.

"Yea," Paula answers.

"You do?" Hopkins responds, apparently surprised by the response. "You said he was a bad guy."

"Yea," Paula says again.

"Do you like bad guys?"

"Yea," says the toddler.

For viewers who failed to pick up the nuances, "Nova" asks Elizabeth Lotus, professor of psychology at the University of Washington, to comment on Hopkins's interviewing techniques. Lotus suggests that viewers could be watching a UFO-abduction memory in the making.

Much of the ground covered by "Nova" is familiar to *Skeptical Inquirer* readers. The program explained how kidnapping by strange creatures has been a common theme in history, with the creatures depending on the culture at the time. It used Committee for the Scientific Investigation of Claims of the Paranormal Fellow Robert Baker to show how false memories can be implanted through hypnosis. It highlighted Loftus's research (she's also a CSICOP Fellow) in which she has found that about one-quarter of the population can be led to embrace memories of events that never happened.

Some of it offered new details.

"Nova" played portions of the infamous "Khrushchev-Kennedy" tape in which Donna Bassett, who infiltrated Mack's group by posing as an abductee, tearfully recalled being aboard a UFO when the two world leaders needed to resolve the unfolding Cuban missile crisis.

In Mack's technique, "there was no skepticism," Bassett now says. "He would believe the most far-fetched things, or at least he seemed to." At the very least, the tape of that hypnosis session suggests that Mack was anxious to pursue the Khrushchev angle as the sometimes-sobbing Bassett insisted that the Soviet leader couldn't possibly be onboard the UFO with her.

BASSETT: He looks like Khrushchev. That can't be.

MACK: Was it Khrushchev?

BASSETT: It's stupid.

MACK: Drop down [the] thirty-five-year-old critical mind for a moment. Did he look like Khrushchev?

BASSETT: Yes.

MACK: Okay. Was anyone else?

BASSETT: There are other people there.

MACK: Anyone else with responsibility like Khrushchev?

BASSETT: Yes. . . . They're happy. They're kissing.

MACK: Who's the other one? What's the other one's name?

BASSETT: The other one. Kennedy. Kennedy. Kennedy.

MACK: You see him?

BASSETT: Yes.

Bassett now says, "The only time he got critical was when I tried to find alternative explanations for some of these experiences myself."

Mack, still puzzled by why Bassett would pose as an abductee, says, "People I know in the experiencer community think she did not hoax. She's an experiencer who never came to terms with her experiences."

Left unspoken is the curious question of why Mack would quote other abductees as the authorities on whether Bassett's story is real. Mack is supposed to be the professional here.

Particularly sobering was the scene where Hopkins was helping a man with a "Comedy Central" T-shirt relive an abduction experience. It helped viewers appreciate how powerful and how disturbingly wrenching these "recollections" can be.

If alien beings aren't kidnapping these folks and these patients would be best treated with reassurances that they are the victims of unusually vivid dreams or hallucinations, the work of people like Mack and Hopkins, who allegedly encourage people to interpret their experiences in the framework of UFO mythology, takes on extraordinarily ominous overtones.

People who claim they are UFO abductees are probably not crazy. By showing far more plausible alternatives and revealing how UFO investigators may be a little too anxious to guide people toward believing they've had an encounter with space creatures, "Nova" has set the standard against which other programs on alien abductions should be measured.

35. NO ALIENS, NO ABDUCTIONS: JUST REGRESSIVE HYPNOSIS, WAKING DREAMS, AND ANTHROPOMORPHISM

Robert A. Baker

Despite earlier efforts to convince the public that regressive hypnosis is a dangerous, unreliable, and deceptive procedure (Baker 1988, 1990), its use both as a standard clinical technique and as a tool used by amateur psychotherapists outside the clinic has grown by leaps and bounds. Both amateurs and professionals have managed to ignore scientific research cautioning the unwary, with the result that over the past twenty years regressive hypnotists have created a wide range of fictions, mistruths, and mythologies that not only mislead and misinform but also have caused a considerable amount of social pain and suffering. Four contemporary legends directly due to the misuse of regressive hypnosis are: UFOs and alien abductions; false memories of childhood sexual molestation; the myth of satanic ritual abuse; and the iatrogenic multiple personality disorders. Each of these social myths has been successfully popularized by a number of best-selling books. Representative are the following: Whitley Strieber's *Communion* (Wilson and Neff 1987), Budd Hopkins' *Intruders: The Incredible Visitations at Copley Woods* (Random House, 1987), Edith Fiore's *Encounters* (Doubleday, 1989), David Jacobs's *Secret Life* (Simon and Schuster, 1992), and John Mack's *Abduction* (Scribner's, 1994), for alien abductions; Ellen Bass and Laura Davis's *The Courage to Heal* (Harper and Row, 1988), Margaret Smith's *Ritual Abuse* (Harper San Francisco, 1993), Judith Herman's *Trauma and Recovery* (BasicBooks, 1992), and Pamela Hudson's *Ritual Child Abuse* (R and E Publishers, 1991), for false memory syndrome; Michelle Smith and Louis Padzer's *Michelle Remembers* (Congdon and Lathes, 1980),

---

This article was written expressly for this volume.

Lauren Stafford's *Satan's Underground* (Harvest House, 1988), Gail Feldman's *Lessons in Evil, Lessons from the Light* (Crown, 1993), and Judith Spencer's *Suffer the Child* (Pocket Books, 1989), for satanic ritual abuse; and Thigpen and Cleckly's *The Three Faces of Eve* (McGraw Hill, 1957), Flora R. Schreiber's *Sybil* (Henry Regnery, 1973), Daniel Keyes's *The Minds of Billy Milligan* (Random House, 1981), and Ralph Allison's *Minds in Many Pieces* (Rawson-Wade, 1980), for alleged multiple personality disorders. All of the false and mythological information found in these books is the result of the use of regressive hypnosis. All of these basically flawed but dramatic and emotionally arousing accounts involve regressive hypnosis as the sole or primary means for obtaining information relevant to their erroneous thesis.

Although disagreement still exists as to whether "hypnosis" is or is not a separate "state of consciousness," there is considerable agreement among all major theorists that "hypnosis" is a situation in which people set aside critical judgment (without abandoning it entirely) and engage in make-believe and fantasy; i.e., they use their imagination (Sarbin and Andersen 1967; Barber 1969; Hilgard 1977; and Spanos and Chaves 1989). Though great differences in the ability to fantasize exist, almost everyone has an imagination and uses it extensively. In recent years many authorities have made imagination a requirement for any successful "hypnotic" performance. Josephine Hilgard (1979) refers to hypnosis as "imaginative involvement," Sarbin and Coe (1972) term it "believed-in imaginings," Spanos and Barber (1974) call it "involvement in suggestion-related imaginings," and Sutcliffe (1961) characterizes the hypnotizable individual as one who is "deluded in a descriptive, nonpejorative sense." Sutcliffe also sees the hypnotic situation as an arena in which people who are skilled at make-believe and fantasy are provided with the opportunity and means to do what they enjoy doing and what they are able to do especially well. In sum, "hypnosis" is mostly and primarily a turning on of the imagination and a royal road to illusion, delusions and fantasyland.

Most psychotherapists, unfortunately, do not read or keep up with the clinical research literature (Campbell 1994), with the result that most so-called clinical truth is, in reality, clinical delusion (Gambrill 1990). Human memory is also seriously flawed and notoriously unreliable—even when people are wide awake. When people are in that highly suggestible state called "hypnosis," memory is even more unreliable and is filled with confabulations, i.e., attempts to fill memory gaps with the most reasonable story possible—not necessarily the truth. Elizabeth Loftus (1993), John Kihlstrom (1993), Richard Ofshe (1993), Nicholas Spanos (1994), and others have recently and brilliantly shown exactly how this happens. Memory is also easily manipulated, as Stephen Ceci and others have demonstrated for the past twenty years (Ceci and Bruck 1993). When "hypnosis" is used on people and they are "regressed," i.e., asked to imagine some past period of their life, and this suggestion is followed up with additional social demands and psychological pressures to comply with the hypnotist's biases and convictions—you will elicit anything and everything but the truth. Reports of being raped, of eating baby's hearts and eyeballs, reports of torture, murder, orgies and sexual molestation, incest—anything that is feared or fearful—are common. The clinician/hypnotist can readily produce "Twilight Zone" or "X-Files" tales and nightmares and anything his client has ever read or imagined or feared may emerge as an actual event or occurrence. Over eight years in the laboratory looking at hypnosis and memory, past lives regressions, future lives *progressions* and other relaxation effects with over five hundred students and community volunteers can be convincing of the fact that it is, indeed, quite easy to persuade people that things that never happened really did happen. Moreover, if the regresses and progresses take the time and trouble or make a sincere effort to account for the sources of their dreams, visions, and imagined scenarios, they usually have little difficulty in doing so.

"Hypnotizing" people is also very very easy. Anyone and everyone can learn to relax people and then use suggestion to create internal images. What you do after the client is relaxed and hanging on your every word, however, is critical. Precisely how the hypnotherapist uses suggestion and exactly *what* he suggests determines whether he elicits the truth (fact) or a confabulation (fiction). People who produce fictional stories are not deliberately lying; they try to remember and cannot. They encounter blank spaces and then they fill in these spaces with plausibilities, not truths. If the therapist believes in Satan and satanic ritual abuses or if he or she believes that all behavioral disorders are due to childhood sexual molestation, or that aliens in UFOs lurk outside our windows every night, then they will find what they are seeking, i.e., another sensational headline, a book, appearances on TV talk shows, and money and attention for themselves and their victims. Many people are so hungry for love, attention, and affection they will even fake diseases and illnesses or physically wound themselves just to get medical attention. Known as "hospital hobos" or as those suffering from "Munchausen's Syndrome" (known technically as *factitious disorders*), such people literally drive physicians up the wall. Psychiatrists Marc Feldman and Charles Ford recently published a fascinating book, *Patient or Pretender: Inside the Strange World of Factitious Disorders* (Wiley 1994), detailing and describing such behavior.

By far the most serious errors made by therapists treating such patients and would-be patients are: (1) buying into the patient's odd belief system; (2) assuming that some horrible trauma in the client's past is responsible for his or her complaint; (3) using regressive hypnosis as a tool for getting "the facts" as well as using it as a method of treatment; (4) believing in the Freudian concept of repression when Holmes (1990) and Pope and Hudson (1994) have convincingly shown that sixty years of experimental research has failed to support its existence; (5) assuming the client's problem will be solved once the repressed material is uncovered and recalled; and (6) perhaps the most serious mistake of all—the therapist's conviction that expressed emotional intensity is an indicator of the validity and truthfulness of the client's memories! Too many sympathetic and empathetic clinicians make the fatal mistake of confusing sobs, tears, moans, and confabulations with the gospel truth. How could anyone carry on like that if what they're remembering were not true? Personally, I have had many individuals put on an Academy Award performance with tears, groans, moans, screams, and hair-tearing from imagined scenarios in a *future* progression. Such emotionality is also fairly routine with people who relive past lives and experience personal tragedies, deaths of loved ones, and other sad events. When conscious and confronted with the tapes of their emotional outbursts, most sheepishly admit they broke down because it all seemed "so very very real and terribly sad." It is amazingly easy to take people too young for World War II back to 1944 and have them relive D day and the channel crossing. If they're particularly imaginative, have seen the movies and read stories of the conflict they will, vicariously, also suffer deeply.

Many people claiming highly unusual and paranormal experiences are, in no manner or way, "crazy" or psychotic. They are, however, deluded; i.e., they do experience false beliefs. While they may have one or two fixed or truly strange ideas, in every other way their behavior is ordinary and normal. Nevertheless, they do make up an ever-growing group of disaffected and alienated people referred to in the past as "eccentric personalities" or people with a "schizotypal personality disorder." While they may be "odd," "spacey," or "strange" they are not schizophrenic. They seldom have any loss of affect, or repeated hallucinations. They manage to get along fine at home, on the job, and in familiar surroundings. Other than being prone to some unusual habits or beliefs, or becoming involved with fringe groups and causes, they seem normal. Typically, they will claim to have received messages from, or to have made contact with, religious entities: angels, demons, Satan, Christ, or even God. They also will claim visions of ghosts or spirits, and they may insist they have supernatural abilities or magical powers, e.g., ESP, clairvoyance, or prophetic accuracy. Many will insist they have a secret mission here on Earth or that they have been sent here from another world to save our planet. They may also report odd perceptions: sensing things that are not present, hearing and seeing things no one else can detect, and communicating with unseen forces or entities. They are often "loners" and "different" because of their fixed beliefs and strange ideas. Few of these people respond well to psychodynamic therapy and their greatest need is the acquisition of social skills. Behavioral treatment by therapists is often quite effective in helping them to deal with day-to-day problems. Therapists should offer comfort and advice without ridiculing their beliefs. Neither medication nor hospitalization is recommended, although this is the first step too many medical personnel usually take. Reassurance and relief from their paranoid fears is the best way to help them stabilize and become socially adjusted. *Increasing their anxieties by supporting their paranoid ideation and beliefs is definitely contraindicated: Regressive hypnosis definitely will not help!* If the therapist uses this approach everything he gets will be colored, twisted, and confabulated. Explaining why the clients' imagination and dreams work the way they do, overcoming their anxieties and fears, and providing comfort and reassurance is the kind of therapy best suited for their social readjustment and eventual recovery.

The major problem with regressive hypnosis is that it turns on the client's imagination, and with the unavoidable demand characteristics of the hypnotic situation plus the suggestions and biases of the therapist as he or she digs into the client's past—anything and everything the therapist unearths will be colored and confabulated so much it will be impossible for either the therapist or the client to know what part of the material is fact and what part is fiction. Once the story is out, however, to the client his or her memories (false as they are) become the only truth possible. This is why so many therapists sincerely believe their clients were sexually molested, satanically abused, and abducted by aliens. How could they make all of this stuff up? Well, let's look at the stories claiming alien abduction as an example.

DIFFICULTIES WITH THE ALIEN ABDUCTION SCENARIO

First of all, as skeptics have long said, "Extraordinary claims demand extraordinary proof." The extraordinary proof in support of people being abducted by aliens has not appeared. In fact, not one scintilla of acceptable, scientifically valid evidence in support of this hypothesis has ever appeared. With regard to this claim, it is not acceptable for UFO and Alien Abduction (AA) believers to make this claim and then demand that the skeptics produce the scientific proof. The burden of proof rests on the shoulders of the claimants—not on the shoulders of the skeptics. Proof of what most skeptics regard as absurd is the job of those who insist that thousands—if not millions—of people are being abducted. Until members of the AA community carry out acceptable scientific research and publish the results in their own *Journal of Experimental UFOlogy & Alien Abductions,* skeptics will continue to insist no acceptable evidence exists in support of these irrational claims. Efforts like those of Dan Wright of MUFON attempting to organize and classify the abduction claims (1994) and Jeff Lindell's work with the autokinetic phenomena (1994) are baby steps in a most welcome and desirable direction.

As for the major difficulties with the AA scenario: first, because of the remarkable similarity of reports from people alleging alien abductions and medical reports from people suffering from sleep paralysis with hypnagogic and hypnopompic hallucinations, it can be safely said that many—if not most—AA claimants are reporting this unusual but not widely known sleep disorder. In fact, the book *Hypnagogia* by Andreas Mavromatis, published in 1987, covers this unique state of consciousness between wakefulness and sleep in great detail. The pulmonary specialist, Dr. Stanley Rehm, has also studied the phenomenon at length (1991, 1993). As far back as 1982, psychiatrist Otto Billig studied the case of three Kentucky women claiming an alien abduction and showed convincingly how the hypnagogic experience, along with regressive hypnosis, convinced them their abduction dream was real (Billig 1982). Numerous reports from the clinic of hypnagogic and hypnopompic dreams and the reports from alleged alien abductees are amazingly alike. Excerpts from Dr. Ronald Siegel's report of his hypnopompic experience, taken from his 1992 book, *Fire in the Brain*:

> I was awakened by the sound of my bedroom door opening. I was on my side and able to see the luminescent dial of the alarm clock. It was 4:20 A.M. I heard footsteps approaching my bed, then heavy breathing. There seemed to be a murky presence in the room. I tried to throw off the covers and get up but I was pinned to the bed. There was a weight on my chest. The more I struggled the more I was unable to move. My heart was pounding. I strained to breathe. . . . Suddenly a shadow fell on the clock. *Omigod! This is no joke!* Something touched my neck and arm. A voice whispered in my ear. . . . In my bedroom I could see only a shadow looming over my bed. I was terrified. . . . I signaled my muscles to move, but the presence immediately exerted all its weight on my chest. . . . *This is no dream! This is really happening!*
>
> A hand grasped my arm and held it tightly. The intruder was doing the reality testing on me: The hand felt cold and dead . . .

> Then part of the mattress next to me caved in. Someone climbed onto the bed! The presence shifted its weight and straddled my body, folding itself along the curve of my back. I heard the bed start to creak. There was a texture of sexual intoxication and terror in the room. . . . (pp. 83–85)

When this hypnopompic report is compared with the alien abduction reports by Strieber in *Communion* (1987) there can be no doubt of their overlap and similarity:

> Sometime during the night I was awakened abruptly by a jab on my shoulder. I came to full consciousness instantly. There were three small people standing beside the bed, their outlines clearly visible in the glow of the burglar-alarm panel. They were wearing blue coveralls and standing absolutely still. . . .
>
> I thought to myself. My God, *I'm completely conscious and they're just standing there.* I thought that I could turn on the light, perhaps even get out of bed. Then I tried to move my hand, thinking to flip the switch on the bedside lamp and see the time.
>
> I can only describe the sensation I felt when I tried to move as like pushing my arm through electrified tar. It took every ounce of attention I possessed to get any movement at all. . . . Simply moving my arm did not work. I had to order the movement, to labor at it. All the while they stood there. . . . I was overcome at this point by terror so fierce and physical that it seemed more biological than psychological. . . . I tried to wake up Anne but my mouth wouldn't open. . . . Again it took an absolute concentration of will . . . but I did manage to smile.
>
> Instantly everything changed. They dashed away with a whoosh and I was plunged almost at once back into sleep. (*Communion,* Avon Edition, 1988, pp. 172–73)

This unusual and interesting phenomenon with both its physiological and psychological effects is summarized in table 1. It should be evident anyone suffering from these physical effects could easily be persuaded something incredible is happening. Such an experiences coupled with with its psychological concomitants and any acquaintance with the AA urban legend would certainly suffice to convince anyone that the little grays are at work again. Table 2 outlines the most common behavioral claims of the AA proponents as well as the most probable explanation for each of these fallacies. Table 3 lists other miscellaneous shortcomings of the abduction scenario.

TABLE 1: SUMMARY OF SLEEP PARALYSIS WITH HYPNAGOGIC AND HYPNOPOMPIC HALLUCINATIONS

A. **Prevalence in General Population:** 4 to 5%

B. **Cause:** Not well known or established, but stress, high anxiety, unusual beliefs, e.g., ghosts, demons, elves, fairies, etc., erratic sleep-wake cycles, poor health, digestive disorders, excessive drug and alcohol use can trigger. These are implicated in some instances but not all.

C. **Predictability:** Low; some individuals may have only one attack in a lifetime. Others may have three or more in a month. One experience also seems to trigger other attacks in the same individual.

D. **History:** Called "Old Hag" attacks in folklore. Known as *Incubus* (male demons attacking sleeping females) or *Succubus* (female demons attacking sleeping males) during Middle Ages. Many literary references.

E. **Associated with:** Cataplexy or partial paralysis and narcolepsy or inability to stay awake, as well as sleep apnea.

I. **Hypnagogic Hallucinations.** Unusually vivid dreams associated with REM sleep. These occur as the individual is falling asleep. Our thoughts, as we are falling asleep, change into dreams. Our disturbed perception makes it difficult to tell whether the perception is real or illusory.

Physiological Effects

Reduced proprioceptive impulses.
Alpha waves (wakefulness) replaced with theta waves (sleep).
Relaxation and muscular paralysis.
Increase in brain activity.
Organic vestibular disturbances.

Psychological Effects

Feeling of floating, falling; clonic spasms or "jerks."
Out-of-body experiences.
Loss of volitional control.
Ishakower phenomena: sparks, flashes or balls of light, faces, animals—all rushing at sleeper; cloudlike masses overwhelming sleeper.
Gritty, salty, or milky sensations around the mouth.
Dizziness, feeling of falling or sinking and distortions of space perception.
Feelings of being overwhelmed or engulfed or manipulated.
Both *sequences* (random images of people objects, and things) and *episodes* (clearly defined themes and schemes, i.e., stories).
Similar to hypnotic state, i.e., suggestibility heightened.

TABLE 1 (CONT.)

II. **Hypnopompic Hallucinations.** These occur when the sleeper is waking up. Many characteristics of hypnagogic state also present.

Physiological Effects

Paralysis of all voluntary muscles including speech.
Rapid heart rate (tachycardia).
Difficult or labored breathing (dyspnea) and restricted or reduced breathing.
Hyperventilation causes reduced oxygen flow to brain. Autoerotic asphyxia.
Hyperacusis, i.e., all sounds are amplified.
Eyes are open and visual sense intact.
Heavy perspiration flow.
Skin temperature and skin resistance changes.
Pupil of the eye dilation.

Psychological Effects

Inability to move or talk.
Vivid visual hallucinations.
Auditory hallucinations.
Extreme fear and feelings of paralysis and helplessness.
Feelings of floating, rising, being moved about.
Feelings of being sexually stimulated or violated.
Feels that he (she) is awake, aware, and conscious.
Dreamer hears, sees, smells, and feels things that are actually present in the environment.
Dreams seem very real, life-like.
Tingling sensations, feelings of cold.
Emotional responses of fear and panic present.
Shadows, distant objects, seen as monsters, ghosts, demons, or aliens.
Extreme fright may cause PTSD in some victims.

III. **Hallucinatory Content.** Determined and influenced by media reports, science-fiction stories, religious beliefs and themes, and UFO abduction stories, as well as current psychological motives, emotions, and needs.

IV. **Commonality of Reports.** All known medical, psychological, clinical, historical and literary accounts of the phenomena are so similar as to be judged identical except for variations in hallucinatory content, i.e., ghosts, elves, demons, fairies, monsters, or extraterrestrial aliens. Particularly striking in terms of reported physiological and psychological effects are the similarities between the experiences of people claiming alien abductions and individuals known to have experienced sleep paxalysis with hypnagogic or hypnopompic hallucinations.

V. **Recommended Reference:** *Hypnagogia,* by Andreas Mavromatis. 1987. London & New York: Routledge & Kegan Paul.

TABLE 2: BEHAVIORAL INDICATORS AND CLAIMS OF ALIEN ABDUCTEES AND THEIR THERAPISTS

Claims	*Most Likely Explanation*
1. Waking up paralyzed with sense of fear of presence in room.	1. Sleep paxalysis and hypnagogic and hypnopompic hallucinations.
2. Experiencing a period of "missing time"—a period of an hour or more in which one cannot remember what happened.	2. Common everyday behavior, wool-gathering, fantasizing, loss of attention, automatic behavior.
3. Feelings of floating or flying without knowing why or how.	3. Muscular relaxation and loss of proprioceptive inputs.
4. Seeing unusual lights, or balls of light or faces, objects or things without knowing what is causing them; hearing buzzing noises, etc.	4. Ishakower phenomenon—part of hypnagogic experience (1938).
5. Having puzzling scars on one's body without memory of how they were caused.	5. Universally common, everyone has them. Nothing unusual.
6. Vague memories of being poked, probed, examined and operated on by huge-eyed aliens.	6. Part of SP and HH experience and due to fear and panic.
7. Memories of sexual operations, e.g., sperm extraction, impregnation, breeding experiments, etc.	7. Reduced oxygen input to brain; similar to autoerotic asphyxiation.
8. Feelings of having small material objects(implants) implanted in various parts of one's body.	8. Part of SP and HH experience; no implants have ever been found.
9. Feelings that aliens communicate via telepathic channels and have communicated vital messages. According to Swords (1989) the 7 most often mentioned are: (1) Colonization; (2) Material gain and power; (3)Threat at home planet; (4) Threat here on Earth; (5) Galactic kinship; (6) Religious conversion; (7) Curiosity & exploration.	9. Partly hallucinatory; hyperacusis also plays a part here. Messages that are communicated are all human motivations—not alien.
10. Under regressive hypnosis skilled therapists can uncover veridical memories of abductions and alien behavior and motivation.	10. Confabulation is the result of all such efforts; hypnosis is a turning on of the imagination. Iatrogenic influences paramount here.

TABLE 3: MISCELLANEOUS SHORTCOMINGS OF THE ALIEN ABDUCTION SCENARIO

1. Number of people being abducted vastly exaggerated.
2. Physical appearance of aliens is identical? No. Many differences.
3. If they appear to be humanoid this proves they are imaginary, due to biological evolutionary evidence and anthropomorphism.
4. Beliefs in aliens mostly inspired by religious and mystical yearnings.
5. Therapists unaware of dangers of regressive hypnosis, e.g., SRAs, false-memory syndrome, MPD, past and future lives work, Snow's *Mass Dreams of the Future* (1989).
6. Unscientific approaches of all AA investigator claims (Wright 1992).
7. Basterfield's "Present at the Abduction" report (1992).
8. Physical absence of all material (credible) evidence to support claims, e.g., no implants, no fetuses, no hybrid children, no souvenirs, no unimpeachable photographs.
9. Most contactees and some abductees "keenly lusted after publicity and profit" (J. Clark 1994).
10. Many alleged abductees tell wild and fantastic stories. Clark calls these "experience anomalies," i.e., while they may be authentic experiences they are in no way either true or valid (1994).
11. Some UFOlogists are more critical of claims than skeptics.

ANTHROPOMORPHISM REVISITED

Perhaps the biggest obstacle of all for the AA believers stems from the well established facts of evolutionary biology. Chris Boyce (1979), E. J. Coffey (1992), Stephen Jay Gould (1990), George Gaylord Simpson (1964), and Frank Tipler (1980), for example, have stressed that the number of chance events that went into the evolution of man is so vast that there is virtually no chance that man as he exists here on Earth exists anywhere else in the universe. Life forms evolving on other planets must go through an equally chance number of events in their development. They, like us, would be similarly unique in the cosmos. What we have in common with any extraterrestrial intelligence is our uniqueness, our singularity. We and all living beings have a profound intimate relationship with our environment.

We did not develop in a vacuum. We are products of our environment. All aliens will have this same sort of symbiotic relationship with their own environments.

As for the alien's physical appearance, it is highly unlikely, most improbable, that they exist as they have been described by the abductees in their small, gray, humanoid form—with no discernable reproductive organs. In a series of fascinating papers concerned with the human tendency to project human qualities upon the external world, E. J. Coffey (1992) reminds us that not only is there no incontrovertible evidence whatsoever that aliens exist, but evolution itself is not the ineluctable following of physical laws but is, instead, merely a chain of contingent events, which easily could have been otherwise. Change any one of the many past events in our biological history—which is a cascade effect—and it will dramatically influence everything that follows. If, by some cruel stroke, the chordates had failed to survive millions of years ago, then neither vertebrates, nor mammals, nor ourselves would ever have evolved. We simply would not be here now. The Burgess shale fossils, representing a time just after the Cambrian explosion 570 million years ago, completely refute the anthropomorphic idea that diversity increased with time. Instead, the evolutionary pattern shows rapid diversification followed by decimation with perhaps as few as 5 percent surviving. In Coffey's words, "The survivors resemble the winners of a lottery rather than creatures better designed than the unlucky majority who do not survive." Stephen Jay Gould (1990) not only concurs, but points out that if we were to replay life's tape there is no reason whatsoever to assume that our particular type of self-conscious being would ever be expected to appear again. As Gould notes, our evolution is not a repeatable occurrence. If anything, we are the embodiment of contingency. What this means is that it is so highly improbable as to approach impossibility that there is any humanoid intelligence of any sort—albeit housed in different bodily frames—to be found anywhere else in the cosmos.

Coffey sums up our anthropomorphic fallacy quite succinctly: "The evolutionary conclusion that humanoid intelligence elsewhere is improbable is not due to any anthropomorphic bias, but it is because of the deep understanding that evolution has no real goal other than adapting creatures to specific local environments. Neither we, nor our mode of intelligence, are the high point of evolution. The pathways of evolution are too circuitous for that to ever be the case" (1992, p. 28). Little gray humanoids who bear a marked resemblance to human fetuses but who are able to communicate telepathically? Dragons, elves, and fairies are equally, if not more, probable. If the aliens came from any space at all it is from the "inner space" inside the human skull rather than "outer space" and the stars. It is also high time we realize that all of our scenarios of extraterrestrial life—from those of the SETI supporters to those of the *Star Trek* series—*all are nothing but projections of ourselves!* If, as the AA believers insist, the aliens and alien technology are in our midst, why would NASA be playing with costly radio telescopes? According to Coffey and many others, the hope for finding human intelligence elsewhere Is a religious conviction: a belief. In his words again: "It is religious in that it rests upon faith—not a rational comprehension of the message the evolutionary record cries out to us, of humans elsewhere there will be none forever" (1992, p. 28).

For many believers in alien visitations and abductions there is no doubt but their convictions are religious and are motivated by the same hopes and longings that motivate SETI scientists. Douglas Curran's beautifully illustrated book, *In Advance of the Landing: Folk Concepts of Outer Space* (1985), shows clearly that religious motives are behind most of the beliefs in saviors in UFOs coming like angels from the skies. The therapist's argument that he must treat the client's *belief* that he was abducted—whether or not they actually were abducted—will not hold water. The belief that one was spirited away when reinforced and confirmed by a therapist not only causes an increase in panic and anxiety—since they're helpless to prevent future abductions—but having the therapist confirm their fear aggravates the initial trauma. If there was ever a smidgen of doubt as to the reality of their prior experience, the therapist's authentication of their abduction removes all traces of conjecture. The result is the helpless client is now likely to have new nightmares about his original experience. Now they are certain their waking-dream was an honest-to-God alien abduction! *Elevating the victim's anxiety level has never been, nor can it ever be in any conceivable way, therapeutic.*

A few weeks ago I was phoned by one of Budd Hopkins's former clients who could not understand why I doubted Hopkins's belief in the abduction delusion. He was quite upset because I challenged the reality of his own experience. After he had the original and powerful emotion of fear revived, reinstated, and then reinforced by Hopkins's hypnotic ministrations, he was now totally convinced his alien experience was real. His last angry letter assured me that I was the one who was crazy, "You weren't there: It didn't happen to you: You just don't know! I know what is *real* and what is not *real*!" The point that he, and every other victim of sleep paralysis and the accompanying hallucinations invariably seem to miss is: *if it did not seem to be very very real it would not be a hallucination!*

In addition to these problems with this sorry modern folktale in which it seems that ABC reporter John Stossel is absolutely correct—"We *are* scaring ourselves to death"—there are even more serious logical flaws in the abductee stories of alien behavior. Most glaring is the obvious lack of any sign of intelligence on the part of the little grays. As for their blatant stupidity, let me count the ways. First, they are scientifically stupid, mathematically stupid, statistically stupid, linguistically stupid, anthropologically stupid, and psychologically stupid. In fact, they are so generally and far-reachingly stupid they reportedly say they want to breed with human beings, clearly the most aggressive and warlike creatures in the cosmos. Humans not only neglect and abuse their young and mistreat the elderly and infirm but are eternally at war with each other and seem totally unable to find peace and contentment even within themselves. A superior intelligence wants to interbreed with this? Aliens are also so naive they think they *can mate with us!* Biologically humans cannot even breed with their closest relatives, the apes and chimpanzees, much less with aliens that, reportedly, have no sex organs. Such things are possible only within the pages of supermarket tabloids, where most of such claims properly belong.

Psychologically, they are so stupid they have no notion whatsoever as to human needs and motives, nor have they betrayed any evidence of empathy or sympathy for human beings. Clearly, they are anatomically and physiologically stupid because in spite of nearly fifty years of collecting human samples they still need more and more of the same things from more and more victims. How much, for alien sakes, is enough? Why don't they ever keep some of the specimens for detailed and intensive study? Why haven't they taken corpses and performed autopsies? Why do they insist on taking their abductees back? Why do they slaughter hundreds of cows when only one or two would give them all the biological information any intelligent scientist would ever need? Unless, of course, they are stupid and unable to profit from past experience. As for the kinds and types of people they abduct, why don't they show some political savvy and abduct the power brokers—heads of state, influential folk like Bob Dole, Rush Limbaugh, or Ross Perot? Or at least a lobbyist, a mayor, or a congressman? If their goal is to truly understand us, why don't they abduct leading scientists, Nobel Prize winners, or at least people who know something worth knowing? If their aim is, as abductees say, breeding, then why don't they abduct Miss Americas, Miss Universes, Olympic athletes, NFL players, or Arnold Schwarzenegger? Or, at the other extremes if they think their own appearance is optimal, then why don't they abduct members of the human family who look most like them, i.e., pygmies? And to learn about human beings they have to kidnap two or three or even four or five million people? To learn about cows, they have to mutilate thousands? Haven't they ever heard of statistics or sampling theory? If their aim is to warn or send messages to humanity, why don't they make use of our wonderfully efficient communication facilities?

Why not show up on TV talk shows or radio? Obviously, they have made no serious or prolonged effort to communicate, to negotiate, or even to announce or explain their presence. As for their clumsy, semi-serious poking and probing of the human anatomy, school kids could do a better job. Moreover, why don't they take an appendix or even a kidney now and then? Of the latter, after all we do have two. Can anyone in his right mind believe that any creature sophisticated enough to cross interstellar space, walk through walls, and communicate telepathically is also totally unable to profit from past experience? If this *is* the case then the human race has nothing whatsoever to fear from this bunch of extraterrestrial idiots. In fact, their intelligence (or lack of it) most resembles that of most of the alleged abductees themselves who seem to have little or no scientific training or understanding and seem to be, generally, in an intellectual fog. Even for those citizens who do appear somewhat lucid and in control of their faculties, their abduction stories certainly resemble dreams and hallucinations more than anything else.

In summary, what we have learned about aliens thus far tells us only one thing: they're not alien at all, they're human. Moreover, they're popped out of the human imagination by human beings with human motives and human needs and desires plus the universal human need, it seems, to believe in entities: in fairies, demons, witches, elves, ETs, and gods as well as in their own egos—their own personal significance and importance in the scheme of things. Their lack of scientific training and experience, analytical thinking, and respect for physical evidence leads much too often to the acceptance of folk beliefs, legends, half-truths, rumor, propaganda, and wishful thinking as substitutes for hard truth and common sense.

Thus far everything alien, when examined closely, also turns out to be very, very human. Stuart E. Guthrie, in his book *Faces in the Clouds* (1993), shows that all religion is anthropomorphic and, like his gods, man also creates his aliens in his own image. If one's vision of human nature is benign then the aliens may well be the gentle, childlike creatures of Spielberg's *Close Encounters of the Third Kind.* On the other hand, if human nature is regarded as beastly and perverse, our aliens may be witches, old hags, hellions, or demons. It is perhaps of some comfort to remember that, according to a recent poll (Gibbs, *Time,* December 27, 1993) 69 percent of the American public not only believe in angels, but 32 percent claim to have made personal contact with them. Given a choice between belief in either angels or aliens, I would pick the angels every time. Not only are they prettier, warmer, gentler, have a better bedside manner and never abduct their hosts, but in many situations the guardian variety have even been known to save lives. Most importantly, they have no desire to interbreed. Well, with only one possible exception that, according to legend, occurred about two thousand years ago. Over the years the entity who resulted from this encounter has brought a considerable amount of comfort and joy to those who believe in his supernatural powers and status. Like all these folk, if you have a need to believe in something extraterrestrial, by all means pick an angel.

REFERENCES

Baker, Robert A. 1988. "The Aliens among Us: Hypnotic Regression Revisited." *Skeptical Inquirer* 12, no. 2: 148–62.

———. 1990. *They Call It Hypnosis.* Amherst, N.Y.: Prometheus Books.

Barber, Theodore X.1969. *Hypnosis: A Scientific Approach.* New York: Van Nostrand, Reinhold.

Basterfield, Keith. 1992. "Present at the Abduction." *International UFO Reporter,* May/June, 13–14.

Billig, Otto. 1982. *Flying Saucers: Magic in the Skies.* Cambridge, Mass.: Schenkman Pub. Co.

Boyce, Chris. 1979. *Extraterrestrial Encounters and Personal Perspective.* Secaucus, N.J.: Chartwell Books.

Campbell, T. W. 1994. "Psychotherapy and Malpractice Exposure." *American Journal of Forensic Psychology* 12, no. 1: 5–41.

Ceci, Steven, and Bruck, M. 1993. *The Child Witness,* Society for Research in Child Development, Social Policy Report, November 1993, Chicago: University of Chicago Press.

Clark, Jerome. 1994. "Big (Space) Brothers." *International UFO Reporter,* March/April, 7–10.

Coffey, E. J. 1992. "The Anthropomorphic Fallacy." *Journal of the British Interplanetary Society*: 23–29.

Curran, Douglas. 1985. *In Advance of the Landing: Folk Concepts of Outer Space.* New York: Abbeville Press.

Gambrill, Eileen. 1990. *Critical Thinking in Clinical Practice.* San Francisco: Jossey-Bass.

Gibbs, Nancy. 1993. "Angels among Us." *Time,* December 27, pp. 56–65.

Gould, Steven Jay. 1990. *Wonderful Life.* London: Hutchinson Radius.

Guthrie, Stuart E. 1993. *Faces in the Clouds.* New York: Oxford University Press.

Hilgard, Ernest R. 1977. *Divided Consciousness: Multiple Controls in Human Thought and Action.* New York; John Wiley and Sons.

Hilgard.,Josephine. 1979. *Personality and Hypnosis: A Study of Imaginative Involvement.* Chicago: University of Chicago Press.

Holmes, David S. 1990. "The Evidence for Repression: An Examination of Sixty Years of Research." In *Repression and Dissociation: Implications for Personality, Theory, Psychopathology and Health,* edited by J. Singer, 85–102. Chicago: University of Chicago Press.

Ishakower, Otto. 1938. "A Contribution to the Psychopathology of Phenomena Associated with Falling Asleep." *International Journal of Psycho-Analysis* 19: 331–45.

Kihlstrom, John F. 1993. "The Recovery of Memory in the Laboratory and Clinic."

Invited paper presented at joint meeting of Rocky Mountain and Western Psychological Associations, April 1993.

Lindell, Jeff A. 1994. "Shooting Venus." *Folklore Forum.*

Loftus, E. F. 1993. "The Reality of Repressed Memories." *American Psychologist* 48: 518–37.

Ofshe, Richard, and E. Watters. 1993. "Making Monsters." *Society* 30, no. 3 (March/April): 4–16.

Pope, H. G., and J. I. Hudson. 1994. "Can Memories of Childhood Sexual Abuse Be Repressed?" Unpublished paper in press. See also "Recovered Memories: Recent Events and Review of Evidence. An Interview with Harrison G. Pope, Jr." *Currents in Affective Illness* 13, no. 7 (July 1994): 5–12.

Rehm, Stanley R. 1991. "Sleep Paralysis and Nocturnal Dyspnea." Paper presented at Advances in Pulmonary and Critical Care Medicine International Symposium, Vienna, Austria, 29–30, 1991. University Kentucky Medical Center, Lexington, Ky.

———. 1993. "Sleep Paralysis in the Salem Witchcraft Trials: A Physiological Basis for 'Spectral Evidence.' " (Forthcoming.)

Sarbin, T. R., and M. L. Andersen. 1967. "Role Theoretical Analysis of Hypnotic Behavior." In *Handbook of Clinical and Experimental Hypnosis,* edited by Jesse E. Gordon. New York: Macmillan Co.

Sarbin, T. R., and W. C. Coe. 1972. *Hypnosis: A Social Psychological Analysis of Influence Communication.* New York: Holt, Rinehart and Winston.

Simpson, George Gaylord. 1964. *This View of Life: The World of an Evolutionist* (chapter 13, "The Nonprevalence of Humanoids"). New York: Harbinger Books.

Snow, Chet B., and Helen Wambach. 1989. *Mass Dreams of the Future.* New York: McGraw-Hill.

Spanos, N. P. 1994. "Multiple Identity Enactments and Multiple Personality Disorder: A Sociocognitive Perspective." *Psychological Bulletin* 116, no. 1: 143–65.

Spanos, N. P., and T. X. Barber. 1974. "Toward a Convergence in Hypnotic Research." *American Psychologist* 29, no. 3: 500–11.

Spanos, N. P., and John F. Chaves, eds. 1989. *Hypnosis: The Cognitive Behavioral Perspective.* Amherst, N.Y.: Prometheus Books.

Sutcliffe, J. P. 1961. " 'Credulous' and 'Skeptical' View of Hypnotic Phenomena." *Journal Abnormal and Social Psychology* 62, no. 2: 189–200.

Swords, Michael D. 1989. "Science and the Extraterrestrial Hypothesis in UFOlogy." *Journal of UFO Studies* 1, no. 1: 67–102.

Tipler, Frank J. 1980. "Extraterrestrial Intelligence Beings Do Not Exist." *Quarterly Journal Royal Astronomical Society* 21, no. 2: 267–80.

Wright, Dan. 1992. "Abductions: Our Dirty Secret." *MUFON UFO Journal,* no. 287 (March 1992): 10–11.

———. 1994. "The Entities: Initial Findings of the Abduction Transcription Project: A MUFON Special Report." *MUFON UFO Journal,* no. 311 (March 1994): 3–7.

Part Six
CROP CIRCLES

36. THE CROP-CIRCLE PHENOMENON

Joe Nickell and John F. Fischer

For years a mysterious phenomenon has been plaguing southern English crop fields. Typically producing swirled, circular depressions in cereal crops, it has left in its wake beleaguered farmers and an astonished populace—not to mention befuddled scientists and would-be "investigators"—all struggling to keep apace with the proliferating occurrences and the equally proliferating claims made about them.

THE MYSTERY AND THE CONTROVERSY

The circles range in diameter from as small as 3 meters (nearly 10 feet) to some 25 meters (approximately 82 feet) or more. In addition to the simple circles that were first reported, there have appeared circles in formations; circles with rings, spurs, and other appurtenances; and yet more complex forms, including "pictographs" and even a crop triangle! While the common depression or "lay" pattern is spiral (either clockwise or counterclockwise), there are radial and even more complex lays (Delgado and Andrews 1989; Meaden 1989; "Field" 1990).

The year 1989 brought no fewer than three books on the cornfield phenomenon and added to the already countless articles on the subject. Soon circles-mystery enthusiasts were being called cereologists (after Ceres, the Roman goddess of vegetation). Circlemania was in full bloom.

By this time, some of the nascent explanations for the early, relatively

---

This article originally appeared in the *Skeptical Inquirer* 17 (Winter 1992). Reprinted with permission.

simple circles had been debunked. The circles' matted pinwheel patterns readily distinguished them from fairy rings (rings of lush growth in lawns and meadows, caused by parasitic fungi) (Delgado and Andrews 1989). The possibility that they were due to the sweeping movements of snared or tethered animals, or rutting deer, seemed precluded by the absence of any tracks or trails of bent or broken stems. And the postulation of helicopters flying upside-down was countered by the observation that such antics would produce, not swirled circles, but crashed 'copters ("England" 1989; Grossman 1990).

A "scientific" explanation was soon attempted by George Terence Meaden, a onetime professor of physics who later took up meteorology as an avocation. In his book, *The Circles Effect and Its Mysteries,* he claims: "Ultimately, it is going to be the theoretical atmospheric physicist who will successfully minister the full and correct answers." Meaden's notion is that the "circles effect" is produced by what he terms the "plasma vortex phenomenon." He defines this as "a spinning mass of air which has accumulated a significant fraction of electrically charged matter." Most evidence, he contends, "suggests that the spinning wind has entered the ionized state known as plasma, and that the vortices are to become plasma balls akin to ball lightning in appearance except that they are much bigger and longer-lived." When the electrically charged, spinning mass strikes a crop field, Meaden thinks, it produces a neat crop circle (1989, 3, 10–11).

Variant forms, he contends, are also allowed by his postulated vortices. For example, of satellite circles Meaden states: "An induction effect may be the consequence of electromagnetic-wave interference resulting in antimodal extrema at the satellite positions, thus leading to secondary rotating plasmas at these locations." Because of their capacity to be ionized, Meaden asserts, the vortices can produce light and sound effects that have been associated with the creation of some circles (Meaden 1989, 60–66).

However, as even one of Meaden's staunchest defenders concedes, "Natural descending vortices . . . are as yet unrecognized by meteorologists" (Fuller 1988). Meaden himself acknowledges that "some from among my professional colleagues who have expressed surprise at the discovery of the circles effect and questioned why it has not previously attracted the attention of scientists, prefer to deny its existence and reject the entire affair as a skillful hoax" (Meaden 1989, 15).

In contrast to Meaden's approach is that of Pat Delgado and Colin Andrews (1989), two engineers who have extensively studied and recorded the crop-circle phenomenon. The pages of their *Circular Evidence* are filled with digressions and irrelevancies—all calculated to foster mystery. For example, we learn of the authors' meeting with a professor and a member of the British Society for Psychical Research, "a meeting," say the authors knowingly, "which we recorded, although the tape was blank when we played it back." A dog that became ill at one site, "some kind of magnetic disturbance" at another, and a plane that crashed after flying over a field where crop circles had appeared some eight weeks before—these are the apparently random occurrences that Delgado and Andrews (1989, 60, 65, 74, 104) attempt to yoke into the mystery.

Overall, Delgado and Andrews hint most strongly at the UFO hypothesis—perhaps not surprisingly, since both have been consultants to *Flying Saucer Review* (Grossman 1990). Although they profess "guarded views" about whether circles and rings have an extraterrestrial source, they frequently give the opposite impression. For example, they go out of their way to observe that a 1976 circle "appeared about seven weeks before a Mrs. [Joyce] Bowles had seen a UFO [and a silver-suited humanoid] just down the road." Again, after visiting one circle Andrews met two teenagers, one of whom had earlier seen "an orange glowing object" nearby. Other mysterious lights and objects are frequently alluded to (Delgado and Andrews 1989, 17, 63, 98).

They rely heavily on anecdotal evidence (i.e., evidence in the form of personal stories, such as the "eyewitness" tales that supposedly prove Elvis Presley still lives). In this respect, it seems that both Meaden and the Delgado and Andrews team have much in common. For example, Meaden offers three alleged eyewitnesses (one an associate of Meaden) to the vortex creation of a circle. Unfortunately, all waited several years before making their claims, and none described the respective events in quite the same way (Meaden 1989, 26–28). Similarly, Delgado and Andrews (1989, 68) associate UFO sightings with circle formation, such as the night "a row of bright lights" was seen at the famous Devil's Punch Bowl at Cheesefoot Head, Hampshire. Use of such conflicting anecdotal evidence prompted one writer to state, rather cynically: "As long as the phenomenon continues, time is on the side of the believers, of course. More and more witnesses, with tales that conform to the dominant myth, will certainly come forward. The witness battalion has already grown since the books appeared" (Shoemaker 1990). Almost predictably, a *hybrid* of the main theories has appeared in "eyewitness" form. Late one evening in early August 1989, or so they claimed, two young men witnessed a circle being formed near Margate, Kent. One of them, a nineteen-year-old, described "a spiraling vortex of flashing light" (a nod to Meaden et al.), which, however, "looked like an upturned satellite TV dish with lots of flashing lights" (a sop to flying-saucer theorists). The youth kept a straight face while posing with the circle for a news photo ("A Witness" 1989–90).

As the crop-circle phenomenon entered the decade of the nineties—bringing with it the emergence of ever more complex forms that earned the sobriquet "pictograms"—the main circular theorists rushed into print their various "Son of Crop Circles" sequels. For example, Paul Fuller and Jenny Randles (who are Meaden's disciples, although, ironically, they are UFOlogists) followed their *Controversy of the Circles with Crop Circles: A Mystery Solved.* Several periodicals devoted to the phenomenon also sprang up, such as *The Cereologist, The Crop Watcher,* and *The Circular,* which was published by the Centre for Crop Circle Studies (Chorost 1991).

If critics of the main theories were not capitalizing on an expanding market of interest in crop circles, they were nevertheless busily poring over the data and pointing out that the prevailing circle theories were, well, full of holes. We were among them.

DATA ANALYSES

Our interest in the swirled-crops phenomenon increased during the latter 1980s. We had already opened a file on the subject, but now we sought to gather information at an accelerated pace. As we studied the incoming data and photographs, we began to formulate hypotheses and to seek out the opinions of agronomists and others who might be helpful. We also enlisted the aid of a computer expert to help us compile and analyze data on the swirled crops—a phenomenon we naturally distinguished from the simple circles or rings that may result from a number of causes (Delgado and Andrews 1989, 160–65).

It soon struck us (as it had many other observers) that the crop-circle phenomenon had a number of potentially revealing characteristics. Cereologists—whether of the "scientific" or "paranormal" stripe—tend either to deny these characteristics or to posit alternative explanations for them. For the implications are serious. While any single attribute may be insufficient to identify a phenomenon, since other phenomena may share that feature, sufficient *multiple* qualities may allow one to rule in or out certain hypotheses so as to make an identification.

The identification we allude to is hoaxing. The characteristics that point to it include an escalation in frequency, the geographic distribution, an increase in complexity over time, and what we call the "shyness effect," as well as a number of lesser features.

An Escalation in Frequency

This aspect of the phenomenon has been well reported. Although there have been reports of circles and rings in earlier years and in various countries—e.g., circles of reeds in Australia in 1966 and a burned circle of grass in Connecticut in 1970—only a few had the flattened swirl feature, and not many of those were well documented at the time (Delgado and Andrews 1989, 179–89; Story 1980, 370–71).

In any case, by the mid-1970s, what are now regarded as "classic" crop circles had begun to appear. In 1976, swirled circles in tall grass were shown near a Swiss village by a man who claimed he was regularly visited by extraterrestrials (Kinder 1987), and Delgado and Andrews (1989) claim an instance in England that same year. When Delgado saw his first circles in 1981, his response was "to share the experience with other people, so I contacted several national papers, along with the BBC and ITN." Then, he says, "Local papers jumped on the bandwagon as soon as they could get the story into print" (Delgado and Andrews 1989, 11–17).

Delgado's use of the word *bandwagon* seems appropriate, since the term refers to an increasingly popular trend or fad. According to Noyes (1989) in an article titled "Circular Arguments," the crop-circles phenomenon "gives the appearance of elaborating and increasing its intrusions from year to year." Writing in 1989, he said that year had "brought more occurrences than ever before."

It was in an attempt to quantify and assess such perceptions that we decided to create a data bank of information on the circles. We used the data in Delgado and Andrews's *Circular Evidence,* which reviewers praised over Meaden's and Fuller and Randles's books for its "level of detail" (Shoemaker 1990) and its being "more comprehensive" (Michell 1989). "Over the years," John Michell (1989) says Delgado and Andrews "have inspected, measured, photographed, mapped and annotated hundreds of circles. . . ." Of course we considered that the incidences of the phenomenon in their book did not represent a complete list, but we intended to look at other sources of data as a cross-check on the sample.

With these caveats in mind we gave a copy of *Circular Evidence* to computer expert Dennis Pearce, an advisory engineer with Lexmark International. Plotting the number of circles per year, Pearce determined, showed a definite (i.e., significantly greater than exponential) increase in the number of crop circles annually from 1981–1987 (Pearce 1991).

This was well supported by data from Meaden's (1989–90) article in *Fortean Times,* "A Note on Observed Frequencies of Occurrence of Circles in British Cornfields." Figures for the four years from 1980 to 1983 were, respectively, 3, 3, 5, and 22; Meaden (1989–90) does not give exact figures for the next few years, but notes they were "rising"; then during the years from 1987 to 1989 the totals went from 73 to 113 to "over 250" annually. For 1990, the figure had again jumped remarkably—to 700 circles in Britain, at least according to Randles (1991a). Small wonder that even moderate voices in the controversy—like Noyes and Michell—insisted the phenomenon was increasing.

However, Meaden and his followers do not accept that there has actually been such a marked increase over the decade. Meaden (1989–90) attributes the increase in part to aerial surveillance begun in 1985. "In addition," he says, "totals for earlier years were still rising, partly as a result of feedback from helpful farmers who were telling us of occurrences of circles on their land which they had known about when they were young." But such anecdotal reports are untrustworthy in the extreme for reasons we have already considered—not the least of which is that we cannot at such a remove distinguish the "classic" phenomenon from other circles and rings.

Meaden mentions even an instance that supposedly occurred in 1678! Widely cited, it is the folk account of an alleged instance of witchcraft in Hertfordshire. Unfortunately for cereologists, the account specifically states that the oats were cut, not bent down in a swirled pattern like the crop circles. Indeed, not only does this story of "The Mowing Devil" (1989–90) fail to support an early historical existence for crop circles, but, says one critic, "the phenomenon's general, if not total, lack of historical precedent is to me its most disturbing aspect" (Shoemaker 1990). Jenny Randles largely agrees with Meaden, although she now increasingly allows for a great number of hoaxes. Speaking of the proliferation of circles that occurred in 1989 and 1990, she states:

> Both those years are hopelessly tainted by the social factors [i.e., the bandwagon effect we mentioned earlier] generated from the huge media hype. Before 1989 we were getting maybe 30 press stories a summer on circles. In 1989 we got up to 300 and in 1990 we had almost 1,000. We can show that this escalation in publicity *preceded* the increase in circle totals and to us this strongly implies a correlation.

In short, she says, the number of circles has been swelled, "heavily contaminated by mass hoaxing inspired by all the publicity" (Randles 1991b).

Here Randles is helping skeptics make their case, because if there is indeed "mass hoaxing," might not the entire phenomenon be similarly caused? We agree with Randles that the escalation of the phenomenon seems to correlate with media coverage of it and that the coverage helped prompt further hoaxes. We provided Dennis Pearce with statistics on crop-circle articles that appeared in the London *Times* from 1986 to 1990, and he specifically commented on "the rapid rise in both locations and number of circles in the years following the London *Times* reports," which, he said, "is to me evidence of human intervention."

Geographic Distribution

A second observed feature of the patterned-crops phenomenon is its predilection for a limited geographic area. As we have seen, prior to the mid-1970s crop circles appeared sporadically at scattered locations in various countries, but since then they have flourished in southern England—in Hampshire, Wiltshire, and nearby counties. It was there that the circles effect captured the world's attention.

In plotting the occurrences of formations among English counties, Pearce confessed that he was "surprised at how localized the phenomenon is." Although there are known exceptions, such as an occurrence in adjacent Wiltshire in 1980, all the pre-1986 cases published by Delgado and Andrews in *Circular Evidence* were in Hampshire, with the vast majority remaining there during the period surveyed (Pearce 1991).

Other sources provide additional evidence for this geographic preference. In 1989 *Time* magazine concluded: "While there have been reports of circles from as far away as the Soviet Union, Japan, and New Zealand, by far the greatest number have appeared in Hampshire and Wiltshire" (Donnelly 1989). The Associated Press, citing a total of 270 circles for the summer of 1989, reported that "two-thirds appeared in a square-mile zone near Avebury in Wiltshire's rural terrain, including 28 in one field" ("England" 1989).

Jenny Randles and Paul Fuller (1990, 50) argue that, to the contrary, "sensible circle researchers have known for some years that circles appear all over Britain, but for various reasons formations appearing away from this area receive little publicity and go unreported outside their local media." But, as Ralph Noyes (1989) counters, many of the reports Randles and Fuller cite are "poorly documented"; he adds that "if credit is given to them all, however, there still remains a strong appearance of an overwhelming concentration of events in Wessex [an area in southern England, named after the old Anglo-Saxon kingdom] and of a phenomenon which has developed explosively in the 1980s."

It may not be coincidence, given the early sporadic circles' association with UFOs, that the clustering occurs where it does. The Wiltshire town of Warminster is "the famous UFO capital of England" (Michell 1989–90); during the sixties and seventies it was what Randles and Fuller term "the center of the UFO universe, drawing spotters from all over the world." Therefore, the area may have provided a unique climate for hoaxes—in the form of UFO-landing sites (as the early Wessex circles were often thought to be)—to flourish.

Some cereologists suggest that, within the densely circle-pocked area, the configurations do have a peculiarity of distribution. Wind-vortex theorists, like Meaden and Fuller (1988), insist that a high proportion of crop circles—some 50 percent—appear at the base of hills, a fact they attribute to the topography's transforming air currents into vortices. However, Delgado and Andrews (1989) observe that the circles form in a variety of locations—"many . . . remote from any hills." On the other hand, it is obvious to skeptics that hills offer good vantage points for hoaxers to view their creations.

Looking beyond the Wessex area, just as the popular media's increasing reportage of the cornfield phenomenon appears to have produced an increase in circle totals—as even Jenny Randles concedes—it also correlates well with the spread of the phenomenon elsewhere. In view of just the data in *Circular Evidence,* Dennis Pearce observed that the number of reported geographical locations in England each year grew at a faster than exponential rate. "I would suspect," he said, "that a natural phenomenon would be either consistently localized or consistently spread about, but not spreading rapidly over time." Also, whereas the circles' pre-English distribution was exceedingly sparse, after newspaper and television reports on the phenomenon began to increase in the latter eighties circles began to crop up in significant numbers around the world. For example, in September 1990 two circles appeared in a Missouri sorghum field and were immediately followed by reports in three other fields—one in Missouri and two in Kansas (McGuire and Adler 1990).

About this time they also had begun to appear in significant numbers in Japan and Canada. Although circles had been reported in Canada sporadically since the mid-1970s, they reappeared with a vengeance in the fall of 1990. Soon after circles turned up in Manitoba in August, the *Toronto Globe and Mail* reported they were "appearing almost weekly now across the Prairies" ("Rings" 1990).

Increase in Complexity

A third characteristic of the patterned-crops phenomenon is the tendency of the configurations to become increasingly elaborate over time. Looking first just at the data in *Circular Evidence,* we see a definite trend. Delgado and Andrews (1989) themselves state: "Before the late 1970s it looked as though single circles were all we had to consider; but, as has always been the pattern, and as we have learnt over the years, something, maybe some intelligent level, keeps one or more jumps ahead. . . ." (p. 122). Again they say: "As soon as we think we have solved one peculiarity, the next circle displays an inexplicable variation, as if to say, 'What do you make of it now?' " (p. 12).

A case in point involves Terence Meaden. When he first published The *Circles Effect and Its Mysteries* in mid-June 1989, setting forth his wind-vortex theory, he declared that "single rings around single circles always rotate in a sense opposite to that of the interior," and he explained that this was necessarily the case (p. 96). Yet no sooner had his book been published than a circle appeared with a ring swirled in the same direction (Michell 1989–90)!

Still, Meaden has clung stubbornly to his vortex theory. He merely acknowledges "the amazing discoveries of recent years" and speaks of the "difficulties of interpretation," concluding:

> Much of the supposed "evolution" of the phenomenon—to which some have even ascribed "intelligence"—can therefore be explained rationally by an insufficiency of data. The more complete the archive becomes, the better this will be appreciated. (Meaden, in Noyes 1990, 76, 85)

To a degree, Meaden is correct; there were some moderately complex forms in earlier periods. But the overall evolution of forms *within the Wessex area* still seems well established, and—worldwide—the emergence of the pictograms in 1990 clearly represented a new phase. States Meaden: "Admittedly, 1990 does look to be exceptional, but just because the reasons for this wait to be clarified, it would be fatuous to decree [that] an alien intelligence is at hand" (Meaden, in Noyes 1989, 85).

The pictograms are wildly elaborate forms with a distinctly pictorial appearance. There have been circles with key shapes and clawlike patterns; complex designs, consisting of circles and rings linked by straight bars and having various appendages and other stylized features; and still other configurations, including free-form "tadpole" shapes and even a crop *triangle* ("Field" 1990; Noyes 1990). Small wonder that Delgado and Andrews, as well as others, suspect that the force that is making the designs is being "intelligently manipulated" ("Mystery" 1990).

But few would have underestimated the cereologists' will to believe had they known of the crop-*message* incident of 1987. The message, written in the typical flattened-crops style and with the words all run together, read: "WEARENOTALONE." Delgado told readers of *Flying Saucer Review*: "At first sight it was an obvious hoax, but prolonged study makes me wonder." Of the crop circles, he said: "Maybe these circles are created by alien beings using a force-field unknown to us. They may be manipulating existing Earth energy" (quoted in Randles and Fuller 1990, 18–19). Or the beings may be terrestrial ones, laboring by the sweat of their brows. At least the pictograms enabled Jenny Randles to wake up to the unmistakable evidence that hoaxes were not only occurring but were running rampant. She has admitted (1991a):

> I do not believe that wind vortices created the pictograms, though serious research into that possibility continues. . . . I can think of very good reasons why the pictograms might well be expected, based on our sure knowledge that crop-circle hoaxing was greatly increased from just a few known cases before 1989 to a far higher figure deduced from my own personal site investigations in 1990. I would put the hoaxes to comprise something over 50 percent of the total.

However, Randles still believes that beyond the hoaxes is a genuine, wind-vortex-caused phenomenon, whereas there seems no need to postulate such. If the "experts" like Meaden, Delgado, and Andrews cannot tell the genuine crop circles from hoaxed ones in 50 percent of the cases, one wonders just what the other 50 percent consist of.

Two other aspects of the patterns' complexity are revealing. The first concerns the lay patterns, which, like the configurations themselves, evolved in complexity. Whereas "the swirl lays of previous years had been orderly and depressed neatly," say Delgado and Andrews (1989), 1987 produced more diverse and far more complex lay patterns (p. 126). They were even more elaborate in 1989 (Noyes 1990, 91).

Then there are the crop designs that have been formed in *stages*—not as single vortex strikes or other brief events. For example, a giant three-ringed circle, with satellites, that was photographed on May 19, 1990, was photographed again on May 27, whereupon it had developed an additional outer ring and ten more satellites (Noyes 1990–93). Surely such work-in-progress seems like nothing so much as the effort of industrious hoaxers.

The Shyness Factor

A fourth characteristic of the patterned-crop phenomenon is its avoidance of being observed in action. There is considerable evidence of this fact.

First, there is its nocturnal aspect. Delgado and Andrews (1989, 156) —who appear to have done the most extensive documentation of the phenomenon—state:

> Many . . . confirmations of nighttime creations come from farmers and people living near circle sites. "It wasn't there last night, but I noticed it first thing this morning" has become almost a stock statement. The evidence is overwhelming that circle creations only occur at night.

Randles and Fuller (1990, 53) agree that "most seem to form during the night or in daylight hours around dawn." However, they insist that "several circles are known to have appeared during daylight," although they do not explain. If they are referring to alleged eyewitness accounts of vortex-formed circles, or other anecdotal evidence, they simply fail to make a case. Besides, this concession that the circles phenomenon is *largely* nocturnal is important since that characteristic seems to run counter to the wind-vortex theory.

Not only does the circle-forming mechanism seem to prefer the dark, but it appears to specifically resist being seen, as shown by Colin Andrews's Operation White Crow. This was an eight-night vigil maintained by about 60 cereologists at Cheesefoot Head (a prime circles location) beginning June 12, 1989. Not only did the phenomenon fail to manifest itself in the field under surveillance, but—although there had already been almost a hundred formations that summer, with yet another 170 or so to occur—*not a single circle was reported for the eight-day period anywhere in England!* Then a large circle and ring (the very set that, being swirled in the same direction, seemed to play a joke on Meaden by upsetting his hypothesis) was discovered about 500 yards away *on the very next day* (Noyes 1990, 28; Michell 1989–90, 47–58).

The following year, the cereologists attempted to profit from their mistakes. This time they conducted a "top secret" operation termed Operation Blackbird, which lasted three weeks beginning on July 23, 1990. They took $2 million worth of technical equipment—including infrared night-viewing camera equipment—to an isolated site where they maintained a nighttime vigil. Reuters quoted the irrepressible Colin Andrews as explaining what happened in the early morning hours: "We had many lights, following that a whole complex arrangement of lights doing all sorts of funny things. It's a complex situation. . . . But there is undoubtedly something here for science" (L. Johnson 1991). Pressed by reporters, Andrews denied that his group could have been fooled by a hoax. However, when they and reporters converged on the site they discovered a hastily flattened set of six circles, with a wooden cross and a Ouija board placed at the center of each.

Other Characteristics

Additional features of the "circles effect" are varied and revealing. Although Ralph Noyes insists that "hoax, as a general theory can be consigned to the dustbin of 'explanationism,' " he notes an interesting apparent characteristic of the phenomenon. This is the way an elaborate feature, a segmented-lay pattern, seemed to have been anticipated a few weeks before: In a nearby field had appeared, says Noyes (1990, 30), "what seems to be the first rough sketch of it." He asks, "Was something *practicing* on that earlier occasion?"

The "something" apparently uses a variety of techniques in doing its sketching and drawing. This is evident not only in the general variety of lay patterns (Swirl, radial, etc.), but also in specific details. For example, Delgado and Andrews (1989, 139) explain that one circle "looks as though the floor was first swept around counterclockwise, then the edge was finished off with a thin circular stroke in the opposite direction." Another circle is notable for "serration marks" found on a few stems and leaves (p. 51); another (p. 142) shows a tightly wound center (as if the stems were wound about a post that was then removed); and so on.

We have already indicated the phenomenon's seeming propensity for mischief. As cereologist Archie E. Roy states (in Noyes 1990, 12), "The phenomenon begins to have the look of a large-scale jokester who is leading us by the nose." John Michell (1989–90) mentions the "perversity" of the phenomenon in producing new patterns that invalidate previous hypotheses. And Hilary Evans (in Noyes 1990, 41) observes that "whoever/whatever is responsible for the crop circles shows every sign of playing games with us."

THE HOAX HYPOTHESIS

We believe that, taken together, the characteristics we have described—the escalation in frequency, the geographic distribution, the increase in complexity, the "shyness effect," and other features—are entirely consistent with the work of hoaxers.

That there *are* hoaxed crop circles no one can deny; the question is of the extent of the hoaxing—that is, whether, if all the hoaxes were eliminated, there would still be a residue of genuine circles that would require postulating some hitherto unproved phenomenon, such as wind vortices or extraterrestrial visitations.

We have seen that some of the cereologists' evidence—the alleged eyewitness accounts, the supposed correlation of circle sites with hilly terrain, and some other claims—is at best unproved and unconvincing. But there are additional assertions.

One claim is that tests of grain from crop circles showed a significant difference in "energy levels" from that in non-crop-circle areas. In fact, a prominent cereologist, the Earl of Haddington, submitted "blind" samples for testing to the Spagyrik Laboratory after receiving confirmation from its director that it could indeed detect the different "energy levels." But in a letter to *The Cereologist* Haddington reported: "Days, weeks passed, months passed, with phone calls at regular intervals always given the same reply. "We will put it [the report of the results] in the post tomorrow." After six months Haddington (1991) concluded: "When they are not told which sample came from a Crop Circle and which from a heap of grain in my back yard they are either unable or unwilling to give a result." (Other claims of differences in "energy levels" come from the many cereologists who employ dowsing or "witching" wands and pendulums to detect the mystical forces. Needless to say, such claims remain unproved.)

Many alleged characteristics of the circles are disputed. Is the phenomenon "silent," as some claim, or sometimes accompanied by a "humming or chirping" sound, which does not appear to be very consistent and might be almost anything? Again, consider the row of "detached" and "dancing" lights seen on the same night a circle was formed: Did they indeed represent a UFO "above the trees" or could they have been a line of hoaxers with flashlights on a distant slope? (Michell, in Noyes 1990, 45; Delgado and Andrews 1989, 68)

If the cereologists cannot offer much in the way of positive evidence, they nevertheless make several negative claims, notably that hoaxers cannot produce circles with the qualities of the "genuine" ones. But what are these qualities? When one of us (JN) debated Delgado on a Denver radio program, it was difficult to get a straight answer from him on this issue.

Delgado's main argument was the alleged lack of broken-stemmed plants in the "genuine" formations, a point he and Andrews make repeatedly in *Circular Evidence*. For example, they say of one circle that "the root end of each stem is bent over and pressed down hard with no damage to the plants, which is why they continued to grow and ripen horizontally" (p. 138).

In response, his various equivocations were pointed out: e.g., in one instance "most" of the plants were undamaged (or rather unbroken; some had "serration" marks on them!) (p. 51). His contradictions were also noted. For instance, Andrews states of one crop ring: "Between the two radial splays was a line of buckled plants. Each one was broken at the knuckle along its stem length." Did he regard the formation as a hoax?

No. He only said, as mysteriously as possible, "These collapsed plants appeared to have suffered whiplash damage, possibly caused by opposing forces meeting" (pp. 63–64). In other words, if the plants are unbroken, that is a mystery; if broken, that is another mystery.

It is entirely possible that the circles with broken plants are merely the less skillfully hoaxed ones. We also considered that the moistening effect of dew on plants bent at night might mitigate against breakage, while agronomists we talked with pointed out that from mid-May to early August the English wheat was green and could easily be bent over without breaking—indeed, could only be broken with difficulty (Blitzer 1990; Daugherty 1990).

Another supposed impossibility is for hoaxers to produce circles without leaving tracks—there allegedly being none in the case of "genuine" circles. But a study of numerous crop-circle photographs in the various publications reveals that virtually every circle would have been accessible by the tractor "tramlines" that mark the fields in closely spaced, parallel rows. In any case, one can carefully pick one's way through a field without leaving apparent tracks.

Can cereologists really tell a hoaxed circle from a "genuine" one? Randles and Fuller (1990, 72) provide a chart of features that supposedly differentiate one from the other. But Randles's belief that the pictograms are not authentic, when Fuller and others think otherwise, suggests that there is little objective basis for making a judgment. Says Dennis Stacy (1991), writing in the *MUFON UFO Journal,* "Most times, we're simply left to take the investigator's word for it, as if some sort of inherited sixth sense were at work."

In fact, rushing to judgment seems a habit of certain cereologists. When ninety-eight circles appeared atop two hills in Wales in less than a week, Colin Andrews was described as spokesman for a "team of top scientists" who were going to investigate. Andrews asserted: "We believe we have something of major proportions. . . . Because of the scale of the formations, we are sure there is no human involvement." Alas, a follow-up report stated: "Red-faced scientists who investigated the ninety-eight mystery circles in the Black Mountains of Wales have discovered they were made by a local farmer—to encourage grouse to settle" (quoted in Randles and Fuller 1990, 97).

In what amounted to a test of his ability, Colin Andrews was asked by a BBC film crew to examine a circle they said they had found. Reportedly, upon visiting the pattern Andrews declared it genuine, but when the BBC explained that it was a hoaxed circle made especially for the occasion, he decided that the circle looked "too perfect" to be genuine after all (Sullivan 1991).

In several cases hoaxers have come forth and confessed, although often the reaction of cereologists is to doubt them. But Jenny Randles and Paul Fuller give credence to the claim of four farmhands from Cornwall that they had created the second 1986 circle-and-ring formation at Cheesefoot Head—one accepted as genuine by Delgado and Andrews (Randles and Fuller 1990, 64–65, 69).

Since we do not have definite knowledge of how particular "genuine" circles were made, the claim that hoaxers cannot have made them is a logical fallacy known as an argument *ad ignorantiam.* As observers like Dennis Stacy (1991) point out, it is now clear that crop circles are comparatively easy to hoax. In fact, the different lay patterns and other details discussed earlier suggest there are various ways of making circles, and indeed various techniques have been described in newspaper articles and books and even detailed by the hoaxers themselves.

An effective method was filmed by the BBC and has impressed those who have seen it. The BBC brought in "a young farmers' tug-of-war team," with the announcer sagely noting that they looked surprisingly practiced as they made their way down the tracks left by the sprayers [i.e., down the tramlines] to the spot we'd chosen." Using a rope to establish a radius, they linked arms, tramped around, and in no time at all had produced what appeared to be a fine circle, which one of them finished off by careful grooming on hands and knees. The announcer commented: "It was roughly at this point that serious doubts crept in and all sorts of little green men were replaced by images of large ruddy ones."

Challenged by the announcer, "Looks as if you'd done it before," the members of the team exchanged grins and knowing looks. Finally, one said, "Well, that would be telling. It's a trade secret. I wouldn't like to say. I think many farmers in Hampshire would be knocking on our doors."

More recently, in September 1991, two "jovial con men in their sixties" claimed they had been responsible for many of the giant wheat-field patterns made over the years. In support of their claim, they fooled Delgado, who declared a pattern they had produced for the tabloid *Today* to be authentic; he said it was of a type no hoaxer could have made. The men said their equipment consisted of "two wooden boards, a piece of string and a bizarre sighting device attached to a baseball cap" (Schmidt 1991). They demonstrated the technique for television crews, e.g., on ABC-TV's "Good Morning America," September 10, 1991, and their proclaimed hoax was publicized worldwide.

The burden remains with the cereologists to justify postulating anything other than such hoaxes for the mystery circles. We feel their time would be better spent attempting to identify more of the hoaxers and to learn what motivates them to do their work. In a chapter titled "Theories

Update" in his and Colin Andrews's *Crop Circles: The Latest Evidence* (1990), Pat Delgado promises cryptically, "What the energy is and who controls it will be explained at a time considered to be more fitting." In the meantime, an insightful reviewer has characterized the circles' effect as "a form of graffiti on the blank wall of southern England" (J. Johnson 1991).

Although the phenomenon has clearly exhibited aspects of social contagion like other fads and crazes—the goldfish-swallowing contest of 1939 comes to mind (Sann 1967, 789–92)—the graffiti analogy is especially apt. Just as graffiti is a largely clandestine activity produced by a variety of scribblers and sketchers possessed of tendencies to indulge in mischief, urge religious fervor, provide social commentary, show off elaborate artistic skills, or the like, so the crop-circles phenomenon has seemingly tapped the varied motives of equally varied circle-makers—from bored or mischievous farmhands to UFO buffs and New Age mystics, to self-styled crop artists, and possibly to others. The phenomenon is indeed mysterious, but the mystery may be only the ever-present one of human behavior.

ACKNOWLEDGMENTS

We are grateful to the following people for their help with this project: Dennis Pearce, for his computer analyses; Morris J. Blitzer and Charles T. Daugherty, for their professional opinions; Barry Karr, Lynda Harwood, and Kendrick Frazier of CSICOP for their research assistance and encouragement; J. Porter Henry, Jr., for transcribing the videotape of a BBC program; Joseph-Beth Booksellers, Lexington, Ky., for help in obtaining books; and to the following, who provided assistance in various ways: Becky Long (Tucker, Ga.), Keith Pickering (Watertown, Minn.), Dixon J. Wrapp (Sebastopol, Calif.), Rob Aken (King Library, Univ. of Kentucky, Lexington), Glenn Taylor, Robert H. van Outer, and Robert A. Baker (Lexington, Ky.), Janet and Doug Fetherling (Toronto, Canada), Christopher D. Allan (Stoke-on-Trent, England), Mr. and Mrs. Charles Hall (Southampton, England), and to all others who have lent assistance, advice, and encouragement.

REFERENCES

BBC program, "Country File." 1988. October 9.

Blitzer, Morris J. 1990. Interview by Joe Nickell, August 28.

Chorost, Michael. 1991. "Circles of Note: A Continuing Bibliography." *MUFON UFO Journal* 276 (April): 14–17.

Daugherty, Charles T. 1990. Interview by Joe Nickell, August 28.

Delgado, Pat, and Colin Andrews. 1989. *Circular Evidence*. Grand Rapids, Mich.: Phanes Press.

Donnelly, Sally B. 1989. "Going Forever Around on Circles." *Time,* September 11, p. 12.

"England Perplexed by Crop-field Rings." 1989. *Denver Post,* October 29.

"Field of Dreams." 1990. *Omni (*December): 62–67.
Fuller, Paul. 1988. "Mystery Circles: Myth in the Making." *International UFO Reporter* (May/June): 4–8.
Grossman, Wendy. 1990. "Crop Circles Create Rounds of Confusion." *Skeptical Inquirer* 14: 117–18.
Haddington, Earl of. 1991. Letter to *The Cereologist* (Spring); quoted in *Skeptics UFO Newsletter,* 10, no. 7, July 1991.
Johnson, Jerold R. 1991. "Pretty Pictures." *MUFON UFO Journal* 275 (March): 18.
Johnson, Larry F. 1991. "Crop Circles." *Georgia Skeptic* 4, no. 3: n.p.
Kinder, Gary. 1987. *Light Years.* New York: Atlantic Monthly.
McGuire, Donna, and Eric Adler. 1990. "More Puzzling Circles Found in Fields." *Kansas City* (Missouri) *Star,* September 21.
Meaden, George Terence. 1989. *The Circles Effect and Its Mystery.* Bradford-on-Avon, Wiltshire: Artetech.
Meaden, G. Terence. 1989–90. "A Note on Observed Frequencies of Occurrence of Circles in British Cornfields." *Fortean Times* 53 (Winter): 52–53.
Michell, John. 1989. "The Alien Corn." *The Spectator* (August 12): 21.
———. 1989–90. "Quarrels and Calamities of the Cereologists." *Fortean Times* 53 (Winter): 42–48.
"The Mowing Devil." 1989–90. *Fortean Times* 53 (Winter): 38–39.
"Mystery circles in British Cornfields Throw a Curve to Puzzled Scientist." 1990. *Newark Star-Ledger,* January 10.
Noyes, Ralph. 1989. "Circular Arguments." *MUFON UFO Journal* 258 (October): 16–18.
———, ed. 1990. *The Crop Circle Enigma.* Bath, England: Gateway Books.
Pearce, Dennis. 1991. Report to Joe Nickell, July 21.
Pickering, Keith. 1990. Unpublished monograph, December 3.
Randles, Jenny. 1991a. "Nature's Crop Circles, Nature's UFO's." *International UFO Reporter* (May/June): 14–16, 24.
———. 1991b. "Measuring the Circles." *Strange Magazine* 7 (April): 24–27.
Randles, Jenny, and Paul Fuller. 1990. *Crop Circles: A Mystery Solved.* London: Robert Hale.
"Rings Mysteriously Appear in Wheat Fields." 1990 *Toronto Globe and Mail,* September 25.
Sann, Paul. 1967. *Fads, Follies and Delusions of the American People.* New York: Bonanza Books.
Schmidt, William E. 1991. "Two 'Jovial Con Men' Take Credit (?) for Crop Circles." *New York Times,* September 10.
Shoemaker, Michael T. 1990. "Measuring the Circles." *Strange Magazine* 6: 32–35, 56–57.
Stacy, Dennis. 1991. "Hoaxes and a Whole Lot More." *MUFON UFO Journal* 277 (May): 9–11, 15.
Story, Ronald D. 1980. *The Encyclopedia of UFO's.* Garden City, N.Y.: Doubleday.
Sullivan, Mike. 1991. "MUFON's Circular Reasoning." *North Texas Skeptic* 5, no. 3: 1–3, 8.
"A Witness from Whitness." 1989–90. *Fortean Times* 53: 37, Winter.

37. CROP-CIRCLE MANIA WANES

Joe Nickell

With England's current crop of grain ("corn," in British parlance), there has sprouted yet another generation of "crop circles"—the swirled-grain phenomenon that for a dozen years had been increasingly imprinted across the southern part of the country and caught the world's attention. "Circlemania" is in decline, however, as I learned on a recent trip to witness the phenomenon first hand.

The giant graffiti began to be noticed about 1980. Notions about their cause ranged from extraterrestrial visitations to wind vortices. However, a study of the features that characterized the phenomenon suggested another answer (Nickell and Fisher 1992). There was an escalation in frequency year after year, a similar increase in complexity (from simple swirled circles in the early period to elaborate pictograms in later years), a noteworthy geographic distribution (the circles' predilection for southern England), and "the shyness factor" (the phenomenon's avoidance of being observed in action).

These characteristics suggested hoaxing as the cause, and, indeed, in September 1991, two "jovial con men in their sixties"—Doug Bower and Dave Chorley—claimed responsibility for many of the crop patterns produced over the years (Nickell and Fisher 1992).

Last summer in London, where I was a speaker and panelist at the *Fortean Times* magazine's "UnConvention," I met various "cereologists" and crop-circle authors—notably Jenny Randles and Paul Fuller—as well as several hoaxers and swirled-crop artists. Fuller introduced himself and

---

This article originally appeared in the *Skeptical Inquirer* 19 (May/June 1995). Reprinted with permission.

announced cheerfully that he had reversed his former position: he now believed only a very small percentage of the grain designs are genuine, those he thought probably being due to the wind vortices postulated by Terence Meaden (1989). In his convention talk, Fuller provided anecdotal evidence that the "genuine" phenomenon had, sporadically, preceded the era of hoaxed circles. Skeptics in the audience, however, were unconvinced by Fuller's data.

To relate briefly some new developments, the incidence of crop circles and pictograms has declined since Doug and Dave (as everyone calls them) confessed their nocturnal activity. Also John Macnish, who as producer of a BBC program and a later bestselling video originally promoted the crop-circle phenomenon, has since written a debunking book, *Crop-circle Apocalypse* (1993), that competes with Jim Schnabel's *Round in Circles* (1993). In addition, some cereologists have become disillusioned and have given up pursuit of the elusive "genuine" phenomenon. These include Meaden (who, I am told, has turned to his amateur archaeological interests) and Pat Delgado—the "father of cerealogy"—who reportedly "'washed his hands' of the subject" and suspended publication of his newsletter. However, he has supposedly since resumed conducting research on crop circles (Fuller 1994a).

Certainly many others continue their interest, which Fuller caters to with his informative periodical *The Crop Watcher,* although he cut back publication from bimonthly to quarterly. Interest among dowsers appears to remain particularly high (more on that in a moment). However, the serious British news media have all but declared a moratorium on the subject. States Fuller (1992b): "Over the past couple of years crop circles have taken a real beating from the skeptics. In my opinion this was long overdue and deserved. Perhaps the lack of attention will drive away the hoaxers."

Some new information pertaining to old matters has also come to light, including Doug Bower's own photos of crop circles he produced at Cheesefoot Head and at Westbury during the early 1980s (Macnish 1993, plates 15–18). Also Macnish (1993) relates how "believers" at a 1992 CSETI watch for alien visitations were fooled by lighted balloons he says were launched by Schnabel and an associate. He also provides new evidence—previously suppressed by cereologists—concerning the elaborate Mandelbrot-set pictogram: discovered at the centers of the design's peripheral circles were small but telltale post holes.

More recent information comes from what I can literally call my field research. On Sunday, June 18, following the close of the Fortean conference, I went on an expedition into the vast wheat crops, conducted especially for me by veteran crop-circle investigators Chris Nash and John Eastmond (both of Southampton University) with an assist from the

United Kingdom's *Skeptical Inquirer* representative, Michael J. Hutchinson (who did not, however, accompany us on the trip). With Chris at the wheel, the three of us motored into the picturesque Wiltshire countryside. We passed through charming thatched-roof villages—including that of Avebury, set amid a great prehistoric circle of standing stones—and came upon a hillside adorned with a giant white horse (one of several ancient effigies formed by exposing the underlying chalk).

By nightfall, we had discovered a handful of circles and pictograms. Two that were reasonably accessible are shown in the accompanying photos. The first was composed of a line of circles—a dozen by my count, or, as Chris waggishly clarified, mocking the exaggerating tone of crop-circle enthusiasts, "exactly a dozen." (Rather than follow the tractor "tram-lines" into the figure, we took a shortcut—carefully picking our way through the wheat.)

It is of course easier to see the overall pattern on a slope from a distance rather than from within the pictogram. The skeptics did not have with them their pole-mounted camera, but John bravely climbed atop my shoulders for a better view and a snapshot from my camera. Examining the swirl pattern, Chris thought the figure a rather ordinary example of a relatively simple pictogram.

The second one we examined was more unusual, with a crescent-and-circle design, but it appeared somewhat older, since the wheat was recovering from having been matted. Amusingly, the farmer had placed crude signs at the gate, requesting that visitors please use the footpath so as not to damage the crop and announcing huffily: "The Circle—It's a Hoax."

Located just opposite the ancient man-made mound, Silbury Hill, the pictogram was nevertheless pronounced genuine by a group of local dowsers who had preceded us to the site. One of them twitched his magical wands for the camera and explained that the swirled patterns were produced by spirits of the earth. He observed that the figure was on a "ley line" (a supposed path of mystical energy) that ran from nearby West Kennet Long Barrow through Silbury Hill to another ancient site. (Chris and I did our best to keep straight faces while the gentleman measured our invisible "auras." After we had compliantly meditated for a few moments, the witching rods indicated our energy fields had expanded from a few inches to several feet. "Wow!" we said.)

Subsequently we made our way to the top of the hill to the nearby ancient barrow, where we encountered a group of young Christian evangelists. As we explored the barrow's tunnel-like passage with its flanking burial niches, overhead the young people sang and rhythmically clapped their hands to "bless" the site and counter any evil forces. Off in the distance was another hillslope adorned with a large pictogram.

After dark we rested over refreshments at an old stone tavern, where cereologists had once congregated in droves. It was now hosting, among others, a group of jockeys and three skeptics—at least one of whom was tired but delighted with the afternoon's rich and colorful experiences. A train ride back to London, arriving at Paddington Station just past midnight, brought to a close my crop-circle adventure.

REFERENCES

Fuller, Paul. 1994a. "John Macnish at the British UFO Research Association, 4th December 1993." *The Crop Watcher* 21 (Spring): 12, 14.

———. 1994b. Editorial. *The Crop Watcher* 21 (Spring): 2.

Macnish, John. 1993. *Cropcircle Apocalypse: Personal Investigation into the Crop Circle Phenomenon.* Ludlow, England: Circlevision. (Reviewed by Paul Fuller, *The Crop Watcher* 21: 18–28.)

Meaden, George Terence. 1989. *The Circles Effect and Its Mystery.* Bradford-on-Avon, Wiltshire: Artetech.

Nickell, Joe, and John F. Fischer. 1992. "The Crop-circle Phenomenon: An Investigative Report." *Skeptical Inquirer* 16: 136–49.

Schnabel, Jim. 1993. *Round in Circles: Physicists, Poltergeists, Pranksters and the Secret History of Cropwatchers.* London: Hamish Hamilton Ltd.

38. LEVENGOOD'S CROP-CIRCLE PLANT RESEARCH

Joe Nickell

In several technical papers, W. C. Levengood purports to show that "Plants from crop formations display anatomical alterations which cannot be accounted for by assuming the formations are hoaxes."[1] Unfortunately, there are serious objections to Levengood's approach. First of all, while he uses various control plants for his experiments, nowhere in the papers I reviewed (see notes 1–4) is there any mention of the work being conducted in double-blind manner so as to minimize the effects of experimenter bias. (As one "cereologist," the Earl of Haddington, said of another laboratory that claimed to detect different "energy levels" between crop-circle and non-crop-circle areas [a concept that appears to have begun with dowsers], "When they are not told which sample came from a Crop Circle and which from a heap of grain in my back yard they are either unable or unwilling to give a result."[5])

The question of bias is important since Levengood's attitudes and assumptions reveal him as a partisan crop-circle "believer" of the Terence Meaden, ion-plasma-vortex variety. Alas, Meaden—who wrote several articles and books advocating the vortex hypothesis—was increasingly forced to conclude that great numbers of crop circles, especially the elaborate pictograms, were produced by hoaxers, and he reportedly abandoned interest in the subject.[6] Levengood's colleague, John A. Burke, seems particularly defiant toward "alleged hoaxers,"[7] as if there were not powerful evidence that most—probably all—of the crop patterns were man-made.[8]

This article originally appeared in the *Skeptical Briefs* 6 (June 1996). Reprinted with permission.

There is, in fact, no satisfactory evidence that a single "genuine" (i.e., vortex-produced) crop-circle exists, so Levengood's reasoning is circular: Although there are no guaranteed genuine formations on which to conduct research, the research supposedly proves the genuineness of the formations. But if Levengood's work were really valid, he would be expected to find that some among the putatively "genuine" formations chosen for research were actually hoaxed ones—especially since even some of Meaden's most ardent defenders admit there are more hoaxed circles than "genuine" ones. (See notes 6 and 8.) In fact, there is now evidence that a major formation that Levengood believes genuine and uses as a basis for theoretical discussion—the "Mandelbrot" formation—was the work of hoaxers.[6]

Although Levengood finds a correlation between "structural and cellular alterations" in plants and their location within crop-circle-type formations (as opposed to those of control plants outside such formations),[1] he should know the maxim that "correlation is not causation." As the noted Temple University mathematician John Allen Paulos recently demonstrated—quite tongue in cheek—there is a direct correlation between children's math ability and shoe size![9] Comments statistician Rand Wilcox of the University of Southern California: "Correlation doesn't tell you anything about causation. But it's a mistake that even researchers make."[9]

That Levengood's work does not go beyond mere correlation in many instances is evident from his frequent concessions: For example, "Taken as an isolated criterion," he says, "node size data cannot be relied upon as a definite verification of a 'genuine' crop formation."[1] Again he admits, "From these observed variations, it is quite evident that [cell wall] pit size alone cannot be used as a validation tool."[1]

Even his alleged correlations are suspect. Citing variations in pit expansion and node size in plants from within the formations, he states: "These energy distributions are by no means uniform."[10] Again, he cites formations where there were increases in plant pit size well outside the formations, saying that "some twenty feet out is the farthest I've seen this energy carryover and so even [though] those crops were standing upright and looked perfectly normal they had been hit." He attributes this to "several different kinds of energy" being involved.[10]

He thus gives the impression that, like Meaden, he is constantly rationalizing new data and attempting to fit it into preconceived vortex notions. Apparently no one has yet independently replicated Levengood's work. One scientist from Colgate did attempt to verify his seed germination claims using some of his seeds but without success.[10] Apparently few mainstream scientists take Levengood's work seriously other than one or two friends who wish "to remain anonymous because of the ridicule.[10]

Until his work is independently replicated by qualified scientists doing "double-blind" studies and otherwise following stringent scientific protocols, there seems no need to take seriously the many dubious claims that Levengood makes, including his similar ones involving plants at alleged "cattle mutilation" sites.[10]

ACKNOWLEDGMENTS

I am grateful to Franklin D. Trumpy, professor of physics, Des Moines Area Community College, for critiquing this article.

NOTES

1. W. C. Levengood, "Anatomical Anomalies in Crop Formation Plants." *Physiologia Plantarum* 92 (1994): 356–63.
2. W. C. Levengood, "Technique for Examining Crop Circle Energetics," Report No. 18 (Pinelandia Lab), October 12, 1993.
3. W. C. Levengood and John A. Burke, "Delineation of Electromagnetic Energy Influencing Crop Formations", Report No. 24, Pinelandia and Am-Tech Labs, September 28, 1994.
4. W. C. Levengood and John A. Burke, "Study of Simulated Crop Formations, 1994," Report No. 27, Pinelandia and Am-Tech Labs, October 10, 1994.
5. The Earl of Haddington, letter to the *Cereologist* (Spring 1991), quoted in the *Skeptics UFO Newsletter* 10 (July 1991): 7.
6. Joe Nickell, "Crop-Circle Mania: An Investigative Update," *Skeptical Inquirer,* in press.
7. John A. Burke, Introduction to W. C. Levengood's Report No. 18 (see note 2).
8. Joe Nickell and John F. Fischer, "The Crop-Circle Phenomenon," chapter 11 of Joe Nickell with John E Fischer, *Mysterious Realms: Probing Paranormal, Historical, and Forensic Enigmas* (Amherst, N.Y.: Prometheus Books, 1992), 177–210.
9. "Statistics Often Misused to Cite Links as Causes, *Lexington Herald-Leader* (Lexington, Ky.), January 5, 1995.
10. W. C. Levengood, telephone interview by A. J. S. Rays, December 8, 1994.

Part Seven
EXTRATERRESTRIAL INTELLIGENCE

39. SEARCHING FOR EXTRATERRESTRIAL INTELLIGENCE

Interview with Thomas R. McDonough

Thomas R. McDonough is an astrophysicist and the coordinator of the Search for Extraterrestrial Intelligence (SETI) for the Planetary Society, the largest space group in the world, with more than 100,000 members on seven continents. Its president is Carl Sagan. McDonough is also a lecturer in engineering at Caltech, a scientific consultant to CSICOP, and the author of several books, including The Search for Extraterrestrial Intelligence: Listening for Life in the Cosmos *and* Space: The Next Twenty-Five Years. *He also writes science fiction on the side, including a novel,* The Architects of Hyperspace. *He was interviewed by Barry Karr, executive director of CSICOP. The interview was first prepared for "The Voice of Inquiry," the new radio magazine of the Center for Inquiry, co-sponsored by CSICOP. It has been revised and updated by Dr. McDonough for the* Skeptical Inquirer.

Q: What exactly is SETI?

McDonough: SETI stands for the "Search for Extraterrestrial Intelligence" and is the name used throughout the world for projects in which scientists are looking for objective, repeatable proof of the existence of civilizations elsewhere in the universe. Usually SETI is understood as the search for radio signals from other civilizations.

Q: How big is SETI, and are any other countries involved in the search?

McDonough: There are three main projects in the United States right

---

This article originally appeared in the *Skeptical Inquirer* 15 (Spring 1991). Reprinted with permission.

now, and there are other countries involved, particularly the Soviet Union and Canada. In Canada, for example, one researcher is now preparing the radiotelescope at the Algonquin Observatory to be used in future search for extraterrestrial intelligence. The Soviet Union has embarked on a project that will look for radio signals. The first of the three main U.S. projects is at Ohio State University and has been operational for about a decade now; it is the longest-running SETI project in the world. The second is the one that NASA is building. It is not yet running, but will probably be operating in 1992 and will then be the world's most powerful SETI project. And finally there are the ones that I am associated with, the Planetary Society's SETI projects at Harvard and in Argentina. These are the most powerful SETI systems currently operating anywhere in the world.

Q: Why do you think that there may be someone, or something, trying to communicate with us?

McDonough: The universe is so big that it seems very unlikely that we could be alone. The mere fact that we have several hundred billion stars in our own Milky Way galaxy, and the fact that this galaxy is just one of perhaps a hundred billion other galaxies, shows the overwhelming number of places in the universe where life could exist. And the idea that we could be alone in the universe seems unlikely just on those grounds alone. But more than that, when you look at the history of life on earth, and at what we're made of, there is nothing that seems terribly unusual when you compare the conditions on the earth with the conditions elsewhere in the universe. For example, the sun is not a weird star, it's not a rare star, it's one of the more common types of stars in the universe. Or if you look at what we're made of—carbon, hydrogen, oxygen, nitrogen—these atoms are not rare. If we were made out of platinum, for example, I might be worried about the basic ingredients for life not being widespread, but we are made out of some of the most common atoms in the universe, ones that we have detected around distant stars and in other galaxies.

Q: Some scientists speculate that an older, more advanced civilization—provided there is one—would have long ago set out to colonize the universe, much like people on earth talk about our plans to someday establish bases on the moon and Mars. The argument goes that, because we haven't seen them, they aren't here, and since they have had billions of years to reach us, that is proof that there aren't any other civilizations out there at all.

McDonough: That is an interesting idea, but it contains an awful lot of hidden assumptions. One is the assumption that a civilization would want to spread like a virus throughout the whole galaxy and colonize every possible piece of real estate in sight. My suspicion is that a really advanced civilization would find so many ways to use its own solar system—the planets, asteroids, moons, and other resources in space—that it wouldn't have as great a need to colonize an entire galaxy. Now, I could be wrong, but the point is that it's crazy, I think, to jump to the conclusion that just because we don't see our own planet being colonized by extraterrestrial beings it means that we are alone in the universe. This is what some people claim.

Q: With the vast amount of noise being generated from the earth, from radio, television, and such, how can you filter through all of that interference? How could you hear a signal if there was one out there?

McDonough: That's a very difficult problem, especially for the NASA system, which looks at all microwave frequencies from 1 to 10 GHz. But in our system we've narrowed down the search. Paul Horowitz, the Harvard scientist who designed and built the system, proposed several years ago that we look for what we call ultra-narrowband signals. What that means is looking for signals that are much narrower than the noise that Mother Nature produces. Nature produces signals that, even in the narrowest cases, are usually spread out over many kilohertz of frequency, very much like a radio broadcast signal on the planet Earth. And so both the artificial signals on Earth and the natural signals of Mother Nature are spread out over a large range of frequencies. What we do is look for very, very narrow signals, signals that are .05 hertz wide, much smaller than even normal radio-station signals. By doing that, we automatically reject most sources of noise, both from nature and from our own civilization. In the Planetary Society SETI programs, we look at eight million frequencies clustered around natural frequencies, such as that of hydrogen (1420 MHz) and hydroxyl (the OH molecule, 1700 MHz). In the NASA system, they will use larger band-widths and cover a much larger range of frequencies. For them it is necessary to develop software to recognize the horrendously complicated noises that our own radio, TV, radar, and satellites produce.

Q: Why do you think a civilization would use those particular frequencies?

McDonough: The best reason would be to avoid the problem of the noisy nature that we live in. The galaxy is chock-full of sources that generate noise. There are gases in space, and magnetic fields, and other things that generate all kinds of noise. In fact, when you tune a radio between stations, part of that hiss that you hear is the hiss from our own galaxy. Or when you look at a television set that's not turned to a local station, then you get snow. And part of that snow is from space. So one of the ways that one civilization could make its signal readily detectable to another civilization is by concentrating all of that energy into the narrowest possible bandwidth, and we calculate that the narrowest practical bandwidth would be around .05 hertz.

Q: How do you know which stars to look at when searching for radio signals?

McDonough: There are two different strategies. The one we use is to search the entire sky. So at Harvard we aim the antenna at one part of the sky, and we let the earth rotate for 24 hours. That allows the antenna to see a strip of sky about half a degree wide during those 24 hour. If we don't see anything, then we raise the antenna another half degree and look at the next strip. The NASA approach is interesting because they are going to combine this kind of technique, which is called the "All Sky Survey," with another type called a "Targeted Search." In a Targeted Search, what you do is look at stars that are similar to the sun, because we think those are the best bets for where life could arise. And so NASA will choose several thousand stars that are much like our sun and focus part of their effort on those stars, and look at a higher level of sensitivity then.

Q: Perhaps we should discuss the differences between SETI and a belief in Unidentified Flying Objects. You believe in the probability of other intelligent life in the universe, but you are critical of reports of UFOs visiting the earth. Why?

McDonough: That's a very sensitive point. The problem is that UFO reports, although interesting to me, are contaminated by so many uncertainties. You don't have a repeatable phenomenon. For example, there has never been a UFO that always came down on the White House lawn at noon every Thursday, which you could go out and check. It is always a case of someone reporting a UFO once, and then it goes away. In the few cases where there have been repetitions, usually we have been able to prove that it was just a natural phenomenon or a hoax. And no one has ever found a piece of material from a supposed UFO landing, a piece of garbage or anything else, that you could take into a laboratory and analyze and show that it was not something from planet Earth.

Q: Would you rule out UFOs visiting here because of the vast distances in space?

McDonough: Well, actually, most of my colleagues do, but I'm not quite that pessimistic. In my books I talk about techniques we can already envision on the drawing board that an advanced civilization could use to travel between the stars. But it's a lot more difficult and a lot more expensive than most people realize. It's certainly not as easy to do as it is in "Star Trek."

Q: Is there a lot of public confusion between SETI and UFOs?

McDonough: There is, and it's one of the headaches we have in this business. Although there have been a few careful UFO investigators, many of them have been very sloppy and very poorly trained, and they have been easily fooled by natural phenomena and hoaxes. So the public tends to think that anyone looking for extraterrestrial beings is automatically involved with UFOs. The scientists who are involved with the SETI project, then sometimes feel contaminated by UFOs, and they don't want to get involved with them because it's such a can of worms. So it has been very difficult for us sometimes to make the public and the Congress understand the difference between the UFO investigations, which are so often flaky, and the SETI investigations, which are supported by many of the leading scientists in the world.

Q: Since all of the polls show that a majority of Americans believe in UFOs, or believe they have seen UFOs, I would think that because of this confusion SETI might be reaping some benefits in public support.

McDonough: Sometimes I think it might, but for the most part we haven't gotten any benefits. Instead we've gotten a lot of problems. The worst one was several years ago when Senator Proxmire killed the entire NASA SETI program because he didn't understand the difference between the flaky people who were too often involved with UFO searches, and the legitimate scientific support for the SETI projects throughout the world.

Q: Did Senator Proxmire give SETI one of his Golden Fleece Awards?

McDonough: Yes, he did. In fact, even before he killed the project, he had given SETI a Golden Fleece Award. And that wasn't enough to satisfy him. He then actually passed a special amendment to the budget that said that NASA could not do anything at all connected with SETI.

Q: That's since been overturned.

McDonough: Right. And in 1990, during the turbulent budget crisis, Congress first killed SETI, but then reinstated it.

Q: Steven Spielberg has probably done a lot to promote belief in UFOs with films like *Close Encounters* and *ET,* but I also know he's a very strong supporter of SETI. It's good to see him devoting time and money to real science, to the real search.

McDonough: Yes, like me, he likes to distinguish between science fiction and science, and he knows the difference. In science fiction, of course, we can go beyond what is known or proved. I was delighted to see him contribute $100,000 to our Planetary Society search at Harvard, enabling us to vastly increase the power of our search.

Q: We actually couldn't converse with extraterrestrials if we did pick up a signal, could we?

McDonough: No, not in the sense of our saying, "Hi, how are you," and their responding, "Well, we're doing pretty good. How are you?" The problem is the enormous distances that exist in the universe. It would mean that, for example, even if they're a light-century away, a hundred light-years away, it would take a hundred years for our signal to get from here to there. And then if they replied, it would take another hundred years for that signal to get back. So it's not the kind of spirited conversation that you would like to have.

Q: Then what would be the benefits of receiving such a signal, of hearing from another civilization?

McDonough: Well, the greatest benefit would just come from knowing that there is another civilization out there. Right now, without that knowledge, we tend to think of ourselves as being unique in the universe, and for people in one culture to think that they are special and very different from people in another culture. So the Russians, the Chinese, and the Americans all think of themselves as being very different from one another. But if suddenly we had positive proof that there were some two-headed little green men in another world with a more advanced civilization, we would begin to see ourselves as a single species in which our differences were trivial compared with the differences between us and them.

Q: Why do you think that if we picked up a signal it would have to be from a more advanced civilization than our own?

McDonough: That's because the civilizations that are not as advanced as we are probably don't have technology we could detect yet. We've been on this planet for millions of years as human beings, but we've only had technology capable of spanning the distances between the stars for the past century. That's when radio was invented. That means that we're not going to be able to detect a civilization at the level of ours at Isaac Newton's time because they didn't have radio. There could be civilizations out there at that level, but we will never detect them with our present technology.

Q: What do you think the message would be? What could they teach us?

McDonough: Oh, I think they could teach us almost everything, because a civilization that is more advanced than we are probably would have solved many of the problems we are now facing: problems of pollution, overpopulation, disease. They may have an encyclopedia that they are just broadcasting at us, which has the answers to all of our questions: how to prevent cancer, how to cure AIDS, and so forth. If we just pick up that information and can decipher it, then we may have a way of jumping ahead thousands of years in our scientific understanding and solving all of our problems.

Q: You said earlier that the NASA project, once it gets under way, could take a decade to finish. What happens if in ten years we don't pick up a signal?

McDonough: I'll be very depressed. What it would mean really is that there won't be any radio signal of the kind we're looking for, but that there could be other signals out there using a technology that we haven't tried yet, or haven't discovered yet. Or it may mean that the signals are not beamed at us, or are not at the frequencies we've used. Most of the thinking so far has gone into detecting signals that are beamed like a lighthouse beacon toward us, because those kind are much easier to detect, even though they are still hard. If they're not beamed at us, if they are just alien "I Love Lucy" shows leaking off into space, then we are going to have to use a much more powerful technology that we don't yet have, and that we can't afford to develop right now. For example, we'd probably have to set up giant radiotelescopes in orbit around the earth, or better yet, on the far side of the moon.

Q: What are the benefits of SETI if we don't ever hear from someone else?

McDonough: Well, if we don't, then the first benefit is that we become convinced that we are alone in the universe, in which case we have to be very careful of our own life. It will make us realize the unique value of our own planet in the universe. And second, it will mean that we will have searched the sky very carefully and probably will have made astronomical discoveries. One of the nice things about SETI is that as you search the sky looking for signals from other civilizations, you automatically discover other objects in the universe, like pulsars, for example.

Q: What is currently exciting? What's currently happening in SETI?

McDonough: The most exciting thing, to me, is the involvement of Argentina in the SETI program. The Planetary Society has an agreement with researchers in Argentina. Two Argentine engineers worked with Paul Horowitz at Harvard building a duplicate of its SETI project. They then took it down to Argentina and installed it there at a radiotelescope. Now we have the full-time observation of the Southern Hemisphere sky. This is the first time that we've systematically studied the southern half of the universe. Up until now, most of the studies in SETI have been in the Northern Hemisphere, and if there is a civilization in the Southern Hemisphere sky we would probably have overlooked it.

Q: Some biologists claim that the probability of intelligence evolving is very small, since out of the billion species that have arisen here only one has achieved our level of intelligence. [See chapter 40 by Zen Faulkes.]

McDonough: Multicellular life has only existed for half a billion years, but our sun will continue to shine without much change for another five billion years. This means that, if we hadn't evolved, billions more species would have arisen.

All it takes is for one species to succeed. Already there are many species on our planet with brains similar in size to ours: dolphins, whales, apes, elephants, and so on. If we hadn't come along, chimpanzees or gorillas might have continued to evolve larger brains. Or maybe bears or some other very different creatures would have succeeded.

As soon as a rudimentary intelligence arises—something comparable to the first tool-using primates—then the smarter ones start outwitting their less-brainy competitors. This means they can have more offspring than the dumber ones. Intelligence gives a creature the ability to defeat its competitors and to survive climate changes.

The game of life is a lottery in which you get billions of chances to play. My suspicion is that, once life evolves on a planet, intelligence has many opportunities to arise during the billions of years the average star shines. Just one success, and they own the place. If I'm right, SETI will probably succeed. If I'm wrong, it will probably fail. That's good science: a testable prediction.

Q: How would you summarize the arguments in favor of SETI?

McDonough: Nobody on earth truly knows whether we're alone or if the universe is buzzing with life. Sticking our heads in the ground never got us anywhere in science. If we don't search, well never find anything. If we do search, the least we'll do is explore the universe. Let's do it!

40. IS INTELLIGENCE INEVITABLE?

Zen Faulkes

Does extraterrestrial intelligence exist? We could search for as close to forever as time allows and never have a definite answer: no matter how long we fail to find a trace of other intelligent life in the galaxy, no matter how long we pile up negative evidence, there will always be other places to look.

Some have observed that this is a discipline whose subject matter has not even been shown to exist, and say that the question of whether extraterrestrial intelligence exists is not even a scientific question. One argument is that the proposition "There is extraterrestrial intelligence" cannot be falsified, and so is not scientific. I disagree. First, while the proposition—"There is extraterrestrial intelligence"—cannot be falsified, the converse—"There is no extraterrestrial intelligence"—jolly well can be. It would be falsified the moment an alien spacecraft landed on Parliament Hill (or the White House lawn, or the steps of the Kremlin). Second, the traditional hallmarks of a good scientific theory are that it is able to predict, control, and explain. Why should it not be fair game to use our current theories to predict the likelihood of alien intelligence and explain why it is probable or not?

The question of extraterrestrial intelligence has certainly gained increased relevance in light of the discovery of planets outside our solar system (e.g., Mayor and Queloz 1995) and the suggestion Mars may have once supported life (McKay et al. 1996). Before dealing with these new developments, a recap of some of the opening gambits in the game of the search for extraterrestrial intelligence (SETI) is in order.

This article first appeared as "Getting Smart about Getting Smarts" in the *Skeptical Inquirer* 15 (Spring 1991), but has been updated and substantially revised for this edition.

Scientific discoveries have repeatedly displaced us from holding a unique position in the cosmos. Earth has been progressively downgraded from the center of the universe to no place special. Biological theories shifted us from God's best and brightest creation to one species among many shaped by evolutionary forces. With such a legacy of scientific thought, brute historical induction made it reasonable to declare, "There is no way that we could be the only intelligent life form in the cosmos." The sentiment is summarized well here:

> The very notion that life is unique to planet earth rings in our minds today like an arrogant aristocentrism, a leftover chauvinism of the sort we recognize in other areas of our thinking and are gradually outgrowing. These inordinate claims to uniqueness and centrality—so understandable before man became informed about himself and his relationship to the natural world—sound like medieval anachronisms. Today such claims just feel wrong, and they no longer compete for intellectual credibility. (Christian 1976)

Unfortunately, claiming that a proposition "feels wrong" never gives it special ontological status.

One useful "lens" that focused debates about extraterrestrial intelligence somewhat is a equation developed by astronomer Frank Drake in 1961. Drake's equation estimates the number of civilizations that we might hope are out there for us to contact. While Drake's equation is formidable looking, the logic is very straightforward.

Start with the number of stars in our galaxy; the raw material, as it were. Call it N_*. As anyone who has ever looked up at a clear night sky will realize, N_* will be a very, very big number. Many people are so impressed with the sheer size of space that they feel that alone is enough to guarantee that intelligent life exists elsewhere. I call this Adams's Argument, because it echoes this sentiment:

> "Space," it says, "is big. Really big. You just won't believe how vastly hugely mindbogglingly big it is. I mean you may think it's a long way down the road to the chemist's, but that's just peanuts compared to space. Listen . . ." and so on. (Adams 1979)

One of the useful things about Drake's equation is that it forces us to look beyond the "space is really big" argument. The size of space notwithstanding, there are other things to think about. For instance, of all the stars, only a fraction (f) of them will have planets. We'll call that f_p. When I wrote an earlier version of this chapter (Faulkes 1991), there were no known extrasolar planets. That, happily, changed with the discovery of 51

Peg, the first known extrasolar planet to orbit a star (Mayor and Queloz 1995), and other extrasolar planets were found in rapid succession (e.g., Butler and Marcy 1996). As of this writing, about eight planets have been found orbiting stars other than our own. As we get a better idea of how many stars do and do not (Walker et al. 1995) have planets, we should have reasonable empirical bases for estimates of fp in the near future.

Of all those planetary systems, only some will have planets or other bodies that could support life, and some could possess several potentially life-supporting worlds. This is not limited to planets, but includes the possibility of life originating on a large moon. Europa, one of Jupiter's moons, has occasionally been nominated as a candidate for having indigenous life. Again, the discovery of other solar systems should eventually allow us to have an empirical estimate of the average number of planets orbiting a star, but for the moment, we'll be content with representing an average number of potentially life-supporting planets per solar system as n_e.

Of all planets capable of supporting life, only a fraction of those will realize that potential and have life arise there. The recent suggestion that Mars may have once supported life (McKay et al. 1996) would, if confirmed, surely be relevant to our estimation of how many planets become biospheres. In the meantime, we'll estimate the probability of life's origin as f_l. Likewise, of all the biospheres that arise, only a fraction of those will develop an intelligent species. I realize that "intelligence" is a bit of a weasel word: it means different things to different people, and the concept tends to falls apart when examined closely. Here, it will denote a species that makes extreme use of tools, language, and numbers. This factor will be shown as f_i, and most of the rest of this paper is devoted to it. Of those smart species, only a fraction of those may have any cultural inclination to build some sort of advanced communications technology. There have certainly been isolationist societies on earth that showed little interest in the affairs of foreigners. Other cultures have not put a high priority on science and technology. This fraction will be f_c.

The final factor to consider is that neither species nor civilizations last forever. This is the variable that has traditionally been considered to be the prickliest of the lot. Considering that people lived for decades with the fear of a nuclear war, and have had that replaced with a prospect of planet-wide environmental collapse, it's not surprising that many people thought that civilizations (including ours) might routinely self-destruct before getting around to a long-term SETI program. We'll express this as the fraction of a planet's life-span that is inhabited by an intelligent species with a technological civilization, and call it f_L.

Put together, Drake's formula is:

$$N = 3D\, N_* f_p\, n_e f_l f_i f_c f_L$$

It is very, very difficult to know what an appropriate estimate for any of these variables is. They've been called, with some disdain but rightfully, "fudge factors." As I noted above, however, we can sharpen our estimates and make them more reasonable as our scientific knowledge increases. The difficulties in estimation aside, if any one of those several fractional values on the right side of the equation were to get very small and approach zero, the rest would be moot.

One of the factors that have not received much attention is the probability of intelligence evolving on a given biosphere, or f_i. Several people have given high estimates for it (Edelson 1978; Russell 1981; Smith 1976), up to and including 1.0: intelligent life always evolves. Smith (1976), for example, claims: "Biological evolution towards greater intelligence naturally occurs as different organisms compete for survival." The notion that intelligence is inevitable once life begins is a fairly strong claim. Echoing the skeptical maxim, "extraordinary claims require extraordinary proof," I suggest that strong claims require principled reasons for making them. Are there any principled reasons, based in biological evolutionary theory, to suggest that intelligence is inevitable, or even likely?

The short answer is no. Evolutionary theory does not predict that there should be any trend to increasing intelligence. For that matter, evolutionary theory does not predict any trend toward any sort of increasing complexity (i.e., more differentiated parts interacting in more ways), which is just a more general version of the intelligence problem. Evolutionary theory says that some animals will survive and more importantly reproduce better in a particular environment than their peers. If those animals are surviving and reproducing better than conspecifics because of a heritable trait, later generations are more likely to possess that trait. There is not a general solution as to how to be evolutionarily successful: there are just too many different environments that change in unpredictable ways all the time. Temperatures rise and fall; sheets of ice advance and retreat; lakes form and dry up; and, occasionally, really big rocks fall out of the sky and into the sea near Mexico. Consequently, organisms react in opportunistic ways, and are not proactive (Gould 1989; Diamond 1990). Significant events in evolutionary history, including both extinctions (Newman 1996) and radiations (Droser et al. 1996), may be records of how organisms are responding to the external environment's pushing and pulling.

One possibility is that while current evolutionary theory does not predict a trend to increasing complexity or intelligence, new theories might one day explain what many perceive to be an increase in complexity that occurred during evolutionary history. Far be it from me to rule out new ideas that might predict increasing biological complexity; there have been candidate theories that purport to predict such trends, but none has gained widespread acceptance. But even the claim that there has been a trend to increasing complexity deserves very careful scrutiny.

We have lived for a very long time with a view of life on this planet that continues to have a powerful grasp on our imagination. It has many names: "the march of progress," "the Great Chain of Being," and *scala naturae.* Bluntly, it is a view that life progresses towards greater complexity and "perfection;" the inevitability of intelligence is a logical corollary. The "march of progress" iconography has permeated our society and popular culture, with humans invariably portrayed as the most complex of all (Gould 1989). It is often unclear, however, how biological complexity is defined, except, perhaps, by how much something resembles us (McShea 1996). For instance, a banner running across the top of the pages of *National Geographic* proclaimed, "The human brain, with its many billions of cells, is the most complex object in the known universe" (Swerdlow 1995). This chestnut seems to be based far more on human vanity than factual data. Why, for instance, is the human brain an "object," whereas a complete human body—which includes the brain and must by definition be more complex—is not? Semantics aside, the brains of many other animals, such as elephants, are larger than ours and probably have many more neurons. Certainly human brains are larger relative to the body than an elephant's, but how is it noticeably more complex? One hopes that by recognizing such biases, we might be able to minimize them.

With due caution and quantitative measures, we can ask, even if a trend toward increasing complexity is not required by evolutionary theory, does the fossil record show such a trend? Perhaps very weakly (McShea 1996). For example, there was a time when there were only unicellular organisms, whereas there are now multicellular organisms. Nonetheless, most of the history of life on this planet is a record of unicellular organisms. To be glib, nothing happened on a macroscopic scale for billions of years (Mayr 1985), even though early life-forms would, by definition, be as simple as possible, with many opportunities to get more complicated.

The origin and diversification of macroscopic, multicellular animals is one instance where a case might be made for increasing complexity. It is clear that multicellular animal life was in full force by the "Cambrian explosion" about 570 million years ago. The "explosion" refers to there being many more fossils found at this point than we have prior to it, so multicellular organisms appear to have arrived in the proverbial blink of an eye (in geologic time). The "sudden" appearance of relatively large animals may be deceiving; trying to decipher the evolutionary events at that time is very difficult because they occurred in the deep past. For example, early animal life may have been small and unlikely to fossilize, so we may be unable to appreciate how complex these animals may have been (McShea 1996; anyone who doubts how complex a microscopic animal can be is advised to look up *rotifers* in an invertebrate biology text). The Cambrian ended in an episode of mass extinction. Certain fossil beds, the most famous of which is the Burgess Shales in Yoho National Park in British Columbia, contain unusually well-preserved remains, including seldom-preserved soft bodies. As Cambrian is the first point in history where we have quite consistent fossil records for macroscopic animals, what were those animals like, and what does it suggest about evolutionary patterns?

The word *weird* hints at what some of these animals were like (Gould 1989). Many of the animals found in the Burgess Shales and similar fossil beds are so unusual that they have been placed in their own high level taxonomic categories, as separate phyla or classes. Gould (1989) argued that while the total numbers of species (the diversity of life) increased since the Cambrian, the number of basic body plans (the disparity of life) actually decreased. The suggestion that the general evolutionary pattern was one of disparity, decimation, and subsequent diversification of the remaining species (Gould 1989) is clearly antagonistic to the notion that there is a trend to intelligence that inevitably culminates in people. There have been several relevant findings since then which weaken the argument somewhat. Several problematic fossils, which were thought to represent entirely new phyla or classes, have been identified as members of contemporary taxa (e.g., Jun-Yuan et al. 1994; Ramskold and Hou 1991). Within the arthropods, quantitative evidence suggests that disparity during the Cambrian was not dramatically different than today (Briggs et al. 1992). Nonetheless, the points remain that very basic body plans were laid down very early, and that extinctions, at least some of which were probably caused by random, nonbiological, external events, have profoundly affected the course of life on this planet. Finally, when one considers that the anomalocaridids were Cambrian-era predators with well-developed eyes, raptorial appendages, and reaching two meters in length (Jun-Yuan et al. 1994), one is hard-pressed to argue how such an animal would be "simpler" than the vast majority of animals alive today. Indeed, nobody has yet to provide any reliable evidence that those lineages that squeaked through episodes of mass extinction were any more complex or "better adapted" than those that died (Gould 1989).

One argument is that increased intelligence is a general purpose solution to cope with an unpredictable and fluctuating environment, and organisms that have it simply must have an advantage over those that do not. This may well be true. There are very few animals in which learning has not been demonstrated. Even tiny nematode worms, with a grand total of 302 neurons in their entire nervous system, can learn and remember (Rankin et al. 1990). Clearly, an elaborate nervous system is not required for an animal to respond adaptively to its environment, although we tend to think that animals with big brains (relative to their body) have greater behavioral sophistication (Bullock 1993). For this discussion, I am going to assume that there is a very general, imprecise correlation between having a more complex and relatively big brain and having a wider range of behavior and increased ability to learn, remember, and solve problems. It is important to note, however, that very few rigorous tests have been made of how those two factors relate (Bullock 1993). Further, even in animals that show abilities we can reasonably describe as some sort of cognition, it is not always obvious as to how those "cognitive" abilities are related to the animals' natural ecology or how they might contribute to reproductive success. Octopuses, for instance, are the only known invertebrates capable of learning by observation (Fiorito and Scotto 1992). This feat is impressive, but it is unclear how it might help octopuses in the wild: octopuses are solitary. Parrots can categorize and understand abstract concepts (Pepperberg 1990a, b), but again, it is an open question as to how the cognitive skills observed in the lab might be used by these birds in more natural settings.

For those who think intelligence must be adaptive, it is encouraging that other animals, not closely related to us, do show these sorts of complicated behaviors. It is also clear that big brains have evolved repeatedly in both vertebrates (Northcutt 1984) and invertebrates (e.g., cephalopods like squid and octopuses). On the face of it, this might suggest that there is not one road of ever-increasing intelligence, but several. On the other hand, there are cases of successful groups of animals whose brains are simpler than those of their less widespread relatives. The brains of hermit crabs and relatives (i.e., Anomura) and of true crabs (i.e., Brachyura) are reduced and simplified compared to other decapod crustaceans, and yet these are tremendously successful animals by any measure you would care to name (Sandeman and Scholtz 1995). Salamanders are another successful group whose brains were apparently simplified before they diverged into a large number of species (Roth et al. 1993). These suggest that the correlation between brain size (and, one supposes, behavioral complexity) and any sort of evolutionary success might be very weak.

One reason why the relationship between evolutionary success and brain size may be weak because any feature in an organism is always a mix of costs and benefits. Even though having a big brain may be advantageous in that it allows for increased cognitive abilities, possible costs related to having a large brain are almost never discussed. It is energetically expensive to maintain a big brain. Our brains consumes about 20 percent of our metabolic energy, compared to about 10 percent for a rhesus monkey or 5 percent for a cat or dog (Armstrong 1990). Another disadvantage of our large brain is that it makes for a large head. This, combined with changes in the structure of the pelvis associated with bipedalism, means that human babies' heads are much larger relative to the birth canal than in other primates. This makes human childbirth much more difficult and hazardous to human females than other primates (Fischman 1994). Having both parent and offspring routinely die during reproduction in long-lived, slowly reproducing mammals like ourselves would not seem to be a surefire recipe for evolutionary success.

There is no particularly pressing reason to think that the origin of technological intelligence on our planet was the end result of a trend or especially probable. On the other hand, it is easy to go too far in the opposite direction and suggest that technological intelligence is spectacularly improbable, as I did in an earlier version of this article (Faulkes 1991). I was overreacting, perhaps, to the popular notion that smart aliens are everywhere. I call this the "crowded Universe" view, and it is emphatically presented in popular media by television shows like "Babylon 5," movies like *Star Wars,* and all their kin and offspring, both greater and lesser. It is suitable, then, that Mars, so often suggested in fiction as home to smart (and, after H. G. Wells, usually hostile) aliens, would be once more the center of attention in this debate.

The suggestion that Mars may have supported life in the past (McKay et al. 1996) is intriguing and exciting. As of this writing, however, it is too early to know what to make of it, and I look forward to seeing how well these results stand up to extensive scrutiny. Confirmation that life once existed on Mars would double biology's sample size, from one biosphere to two. It suggests that the generation of living organisms may occur under a wide range of conditions. There is a more depressing side to the announcement of possible past Martian life, however. Mars may be an entire biosphere that has gone extinct. We find living organisms living and often thriving in our planet's most hostile locations, so terrestrial life appears marvelously tenacious and resilient. There is no evidence of life on Mars now, suggesting that if life originated on the red planet, it never managed to get a toehold: No macroscopic organisms, no increasing complexity, no smart Martians carved out by the forces of natural selection.

This is one reason why, unlike some (Simpson 1964; Mayr 1985), I cannot being myself to end this chapter by saying that SETI is a waste of time, money, and effort. What if Mars is a planet where all life died out? What if we look long and hard for other technological civilizations, and find nothing? It immediately suggests that booming, buzzing, life-infested planets like ours are probably uncommon at best, and that technological civilizations are very rare. "It would speak eloquently of how rare are the living beings of our planet and would underscore, as nothing else in human history has, the individual worth of every human being" (Sagan 1980). We should take into account how probable or improbable alien intelligence may be when deciding how much of our effort SETI should occupy. I liken SETI to playing the lottery: the odds may be long, even unfair, but the payoff is big. The impact on science, especially biology, if we did contact an alien intelligence would be so gargantuan that it seems foolish to abandon the entire affair. As Heinlein (1978) wrote, "Certainly the game is rigged. Don't let that stop you; if you don't bet, you can't win." SETI remains an intellectual game worth playing, even if only on a small scale.

REFERENCES

Adams, D. 1979. *The Hitchhiker's Guide to the Galaxy.* Pan Books: London.

Armstong, E. 1990. "Brains, Bodies and Metabolism." *Brain, Behavior and Evolution* 36: 166–76.

Bullock, T. H. 1993. "How Are More Complex Brains Different? One View and an Agenda for Comparative Neurobiology." *Brain, Behavior and Evolution* 41: 88–96.

Briggs, D. E. G., R. A. Fortey, and M. A. Wills. 1992. "Morphological Disparity in the Cambrian." *Science* 256: 1670–73.

Butler, R. P., and G. W. Marcy. 1996. "A Planet Orbiting 47 Ursae Majoris." *Astrophysical Journal* 464: L153–L156.

Christian, J. 1976. "The Story of Life: Earth's Four-billion-year Beginning." In *Extraterrestrial Intelligence: The First Encounter,* edited by J. Christian, 15–31. Amherst, N.Y.: Prometheus Books.

Diamond, J. 1990. "Alone in a Crowded Universe." *Natural History* (June 1990): 30, 32, 34.

Droser, M. L., R. A. Fortey, and Li Xing. 1996. "The Ordivician Radiation." *American Scientist* 84: 122–31.

Edelson, E. 1978. *Who Goes There?* New York: McGraw-Hill.

Faulkes, Z. 1991. "Getting Smart about Getting Smarts." *Skeptical Inquirer* 15: 263–68.

Fiorit, G., and P. Scotto. 1992. "Observational Learning in Octopus Vulgaris." *Science* 256: 545–47.

Fischman, J. 1994. "Putting a New Spin on the Birth of Human Birth." *Science* 264: 1082–83.

Gould, S. J. 1989. *Wonderful Life.* New York: W. W. Norton.

Heinlein, R. A. 1978. *The Notebooks of Lazarus Long.* New York: G. P. Putnam's Sons.

Jun-yuan, Chen, L. Ramskold, and Zhou Gui-qing. 1994. "Evidence for Monophyly and Arthropod Affinity of Cambrian Giant Predators." *Science* 264: 1304–1308.

Marcy, G. W., and R. P. Butler. 1996. "A Planetary Companion to 70 Virginis." *Astrophysical Journal* 464: L147–L151.

Mayor, M., and D. Queloz. 1995. "A Jupiter-mass Companion to a Solar-type Star." *Nature* 378: 355.
Mayr, E. 1985. "The Probability of Extraterrestrial Intelligent Life." In *Extraterrestrials: Science and Alien Intelligence,* edited by E. Regis, 23–30. Cambridge: Cambridge University Press.
McKay, D. S., E. K. Gibson, Jr., K. L. Thomas-Keprta, H. Vali, C. S. Romanek, S. J. Clemett, X. D. F. Chillier, C. R. Maechling, and R. N. Zare. 1996. "Search for Past Life on Mars: Possible Relic Biogenic Activity in Martian Meteorite ALH84001." *Science* 273: 924–30.
McShea, D. W. 1996. "Metazoan Complexity and Evolution: Is There a Trend?" *Evolution* 50: 477–92.
Newman, M. E. J. 1996. "Self-organized Criticality, Evolution and the Fossil Record." *Proceedings of the Royal Society of London* B 263: 1605–10.
Northcutt, R. G. 1984. "Evolution of the Vertebrate Central Nervous System: Patterns and Processes." *American Zoologist* 24, no. 3: 701–16.
Pepperberg, I. M. 1990a. "Cognition in an African Grey Parrot (Psittacus Erithacus): Further Evidence for Comprehension of Categories and Labels." *Journal of Comparative Psychology* 104: 41–52.
———. 1990b. "Some Cognitive Capacities of an African Grey Parrot (Psittacus Erithacus)." In *Advances in the Study of Behavior,* edited by P. J. B. Slater, J. S. Rosenblatt, and C. Beer, 357–409. New York: Academic Press.
Ramskold, L., and X.-G. Hou. 1991. "New Early Cambrian Animal and Onychophoran Affinities of Enigmatic Metozoans." *Nature* 351: 225–28.
Rankin, C. H., C. Chiba, and C. Beck. 1990. "Caenorhabditis Elegans: A New Model System for the Study of Learning and Memory." *Behavioral Brain Research* 37: 89–92.
Roth, G., K. C. Nishikawa, C. Naujoks-Manteuffel, A. Schmidt, and D. B. Wake. 1993. "Paedomorphosis and Simplification in the Nervous System of Salamanders." *Brain, Behavior and Evolution* 42: 137–70.
Russell, D. 1981. "Speculations on the Evolution of Intelligence in Multicellular Organisms." In *Life in the Universe,* edited by J. Billingham, 259–75. Cambridge: MIT Press.
Sagan, C. 1980. *Cosmos.* New York: Random House.
Sandeman, D., and G. Scholtz. 1995. "Ground Plans, Evolutionary Changes and Homologies in Decapod Crustacean Brains." In *The Nervous System of Invertebrates: An Evolutionary and Comparative Approach,* edited by O. Breidbach and W. Kutsch, 115–38. Basel: Birkhauser Verlag.
Simpson, G. G. 1964. "The Non-prevalence of Humanoids." *Science* 143: 769–75.
Smith, R. 1976. "The Education of Human Intelligence." In *Extraterrestrial Intelligence: The First Encounter,* edited by J. Christian, 147–59. Amherst, N.Y.: Prometheus Books.
Swerdlow, J. L. 1995. "Quiet Miracles of the Brain." *National Geographic* 187, no. 6: 2–41.
Walker, G. A. H., A. R. Walker, A. W. Irwin, A. M. Larson, S. L. S. Yang, and D. C. Richardson. 1995. "A Search for Jupiter-mass Companions to Nearby Stars." *Icarus* 116: 359.

CONTRIBUTORS

ROBERT A. BAKER is professor emeritus of psychology, University of Kentucky.

ROBERT E. BARTHOLOMEW completed his Ph.D. dissertation in June 1989, a study of collective behavior, especially of historical UFO sightings.

JOSEPH A. BAUER is a surgeon in Cleveland, Ohio, and a member of South Shore Skeptics.

WILLIAM BLAKE is an aerospace engineer at the Wright Laboratory, Wright-Patterson Air Force Base.

ROBYN M. DAWES is University Professor in the Department of Social and Decision Sciences, Carnegie Mellon University.

C. EUGENE EMERY, JR., is the science writer for the *Providence Journal.*

ZEN FAULKES is at the Department of Biology at McGill University.

JOHN F. FISCHER is a forensic analyst in a Florida crime laboratory.

Physicist/engineer KINGSTON A. GEORGE retired recently after thirty years of air force civil service and continues as a private aerospace consultant. His initial appointment in 1961 was as an operations research analyst for the 1st Strategic Aerospace Division at Vandenberg AFB, California, where he pioneered many aspects of range safety and range instrumentation systems deployment. As chief engineer for safety at Vandenberg AFB in 1989, he was honored in Washington, D.C., as the recipient of the Air Force Association's Senior Civilian Manager of the Year Award.

PHILIP J. KLASS, a senior editor with *Aviation Week & Space Technology* magazine, is a Founding Fellow of CSICOP and chairman of its UFO Subcommittee. His most recent book on the subject is *UFOs: The Public Deceived* (Prometheus Books).

THOMAS R. MCDONOUGH is an astrophysicist and coordinator of the Search for Extraterrestrial Intelligence (SETI) for the Planetary Society.

MATTHEW MULFORD is at the London School of Economics.

JOE NICKELL is Senior Research Fellow at CSICOP and Investigative Files columnist for *Skeptical Inquirer.*

JAMES OBERG is a computer engineer, the author of many books on space, and a member of CSICOP's UFO Subcommittee.

IAN RIDPATH is a well-known science and space writer and a member of the U.K. branch of CSICOP. He is the author of several books and editor of the *Encyclopedia of Astronomy and Space.*

ROBERT SHEAFFER is a freelance writer and longtime skeptical UFO investigator. He is the author of *The UFO Verdict: Examining the Evidence* (Prometheus Books).

ARMANDO SIMÓN is a psychologist interested in applying psychological research to UFO sightings and beliefs. He is staff psychologist for the Texas Department of Corrections, Rosharon, Texas.

LLOYD K. STIRES is a professor of psychology at Indiana University of Pennsylvania.

TREY STOKES has worked as a creature effects artist for such films as *Species, The Abyss, Batman Returns, RoboCop II,* and *The Blob.*

DAVID E. THOMAS is a physicist. He is vice president and communications officer of New Mexicans for Science and Reason.

RICHARD L. WEAVER, Col. USAF, is director of Security and Special Program Oversight for the air force.

ROBERT R. YOUNG is education chairman of the Astronomical Society of Harrisburg, Pennsylvania.